SPLITTING OF TERMS IN CRYSTALS

Hans A. Bethe

COMPLETE ENGLISH TRANSLATION

[From: Annalen der Physik, Volume 3, pp. 133-206 (1929)]

Printed in the United States Consultants Bureau, Inc. Price: $3.00

PHYSICS

SPLITTING OF TERMS IN CRYSTALS

Hans A. Bethe

(With 8 Figures)

The influence of an electric field of prescribed symmetry (crystalline field) on an atom is treated wave-mechanically. The terms of the atom split up in a way that depends on the symmetry of the field and on the angular momentum l (or j) of the atom. No splitting of s terms occurs, and p terms are not split up in fields of cubic symmetry. For the case in which the individual electrons of the atom can be treated separately (interaction inside the atom turned off) the eigenfunctions of zeroth approximation are stated for every term in the crystal; from these there follows a concentration of the electron density along the symmetry axes of the crystal which is characteristic of the term. - The magnitude of the term splitting is of the order of some hundreds of cm^{-1}. - For tetragonal symmetry, a quantitative measure of the departure from cubic symmetry can be defined, which determines uniquely the most stable arrangement of electrons in the crystal.

§ 1. Introduction.

There are two immediately evident points of departure for the wave mechanics of crystals: on one hand, we may treat the crystal as a single whole, and accordingly describe it in terms of a spatially periodic potential and eigenfunctions of the same spatial periodicity, whose modulation by the more detailed structure of the atoms contained in the crystal is included only in the second approximation. This method has primarily been used by Bloch[1] for the treatment of the electric conductivity, and seems particularly suitable in the case of electrons that are to a considerable degree "free". On the other hand one can start with the free atom and treat its perturbation in the crystal in a way analogous to the method of London and Heitler[2]; one will be able to use this procedure primarily for the calculation of lattice energies, lattice spacings, etc.

There are two reasons to expect a perturbation of the free atom on its inclusion in a crystal: on one hand, the atom will enter into exchange of electrons with the other atoms of the crystal, i.e., its permutation group is changed. This exchange effect will have to be treated in a way quite analogous to the case of a molecule, there being at most a purely quantitative difference in regard to the number of neighbors with which the exchange can take place. On the other hand there acts on the atom in the crystal an electric field arising from the other atoms

[1]F. Bloch, Ztschr. f. Phys. Vol. 52, p. 555 (1929).
[2]F. London and W. Heitler, Ztschr. f. Phys. Vol. 44, p. 455 (1927).

which has a definite symmetry, which removes the directional degeneracy of the free atom. The qualitative difference between the molecule problem and the crystal problem is to be sought in this different spatial symmetry. In this paper, we shall deal with the specific perturbations of the free atom that are occasioned by the symmetry of the crystal, without at first going into the question of the electron-exchange phenomena for the atoms of the crystal; thus, we shall at the same time obtain a starting point for the treatment of these latter phenomena.

An electric field of definite symmetry will cause a splitting of the terms of the unperturbed atom that is analogous to the Stark-effect splitting and will just be characteristic for the symmetry of the field, i.e., for the symmetry of the situation of the atom in the crystal. The number of components into which a term of the free atom is split increases with decreasing symmetry. The amount of the splitting can be of very different orders of magnitude, and in regard to this one will have to distinguish three cases.

1. Stark-effect splitting in the crystalline field large in comparison with the separation of different multiplets: the effect of the crystalline field on the atom overcomes the interaction of the electrons in the atom and in first approximation removes the coupling between them. We then begin with a model of the free atom in which only the occupation numbers of the quantum cells specified by the principal quantum number n_i and the azimuthal quantum number l_i (of the individual electron) are prescribed, and the term splitting corresponding to the exchange degeneracy is left out of account. (The electrostatic interactions of the electrons can be included, say by the Hartree method of the self-consistent field.)

In the first approximation we then consider the perturbation of individual electrons outside closed shells by the field of the other atoms in the crystal, i.e., the possible orientations of the orbital angular momentum l_i of the individual electron with respect to the crystal axes, and the "Stark-effect splitting" of the terms of the atom that is thus produced. In the second approximation, account would have to be taken of the electron exchange inside the atom; this in general brings about a further splitting of the terms, naturally of the order of magnitude of the separation between different multiplets of the free atom. Finally, the interaction between orbital angular momentum and spin, again produces the multiplet splitting in the usual way.

2. Crystal splitting of intermediate size, i.e., small in comparison with the separation of different multiplets, but large in comparison with the differences within a single multiplet. In this (probably the most frequent) case, we must start

with the free atom including the term splitting by exchange degeneracy, but without taking into account the interaction of spin and orbit. Then this atom has to be inserted into the crystal, and we must examine the orientations of its total orbital angular momentum with respect to the axes of the crystal and the corresponding term values, and finally study the interaction between spin and orbit for fixed orientation of the latter in the crystal.

3. Crystal splitting small in comparison with the separations within a multiplet. The quite "finished" atom with account already taken of the interaction between electronic orbits and spins is subjected to the perturbation by the electric field of the crystal. This means that the vector that receives an orientation relative to the crystal axes is the total angular momentum j of the atom, and no longer, as in cases 1 and 2, the orbital angular momentum, because this latter vector remains firmly coupled to the spin, whereas in cases 1 and 2, the coupling was removed by the crystalline field.

We shall use the following quantum numbers to describe the Stark-effect splitting:

λ azimuthal crystal quantum number to specify the orientation of the orbital angular momentum of the atom in the crystal.

λ_i azimuthal crystal quantum number of the individual (ith) electron.

l_i ordinary azimuthal quantum number of the individual electron in the free atom.

μ inner crystal quantum number (orientation of the total angular momentum j of the atom in the crystal).

Since in the case of crystal quantum numbers, we are in general not dealing with "true" quantum numbers, which can be interpreted as angular momentum around an axis or something of that sort, but only with distinctions between different group-theoretical representation properties of the individual terms, we write them not as numbers, but as Greek letters.

An atom is then characterized by the following quantum numbers, in the three cases mentioned above:

1. Large crystal splitting: n_i, l_i, λ_i, λ, μ.
2. Intermediate crystal splitting: n_i, l_i, l, λ, μ.
3. Small crystal splitting: n_i, l_i, l, j, μ.

For the present, we shall treat the three possible cases - orientation of the orbital angular momentum of the individual electron, of the total orbital angular momentum, or of the total angular momentum of the atom in the crystal - together and make group-theoretical calculations of the number of components into which a term splits for given angular momentum (representation of the rotation group) and given symmetry of the situation of the atom in the crystal. The case of the orientation of a half-integral total angular momentum (double-vlaued representation of the rotation group) will have to be treated separately. We shall also indicate the angle-dependent factors of the zero-order eigenfunctions belonging to the individual components resulting from the splitting, which will be required for calculations of the exchange effects. We shall then discuss separately the three cases distinguished by the order of magnitude of the crystal splitting, paying particular attention in this connection to the symmetry of the electron density distribution. Finally we shall calculate the splitting for ionic crystals.

I. Group-Theoretical Solution.

§2. Outline of the Solution

As is well known, the Schrödinger differential equation of the free atom is invariant with respect to arbitrary rotations of the coordinate system and to an inversion in the nucleus (and with respect to exchange of the electrons, which, however, is not of interest to use for the present). Its substitution group[1] includes, with other factors, the rotation group of the sphere, which possesses one irreducible representation of each dimensionality $2\ell + 1$ ($l = 0, 1, 2 \ldots$).

If we now insert the atom into a crystal, then the symmetry of the potential energy is reduced from spherical symmetry to the symmetry of the position that the atom occupies in the crystal, e.g., for insertion into a crystal of the NaCl type to cubic-holohedral symmetry, for the ZnS type to cubic-hemihedral symmetry. The substitution group of the Schrödinger differential equation for the atom in question now includes only such symmetry operations as leave the position of the nucleus of this atom unchanged and at the same time carry the entire crystal over into itself (symmetry group of the crystal atom). The way to deal with such a diminishing of the symmetry has been indicated in general by Wigner. What is required is just to accomplish the reduction of the representation of the original substituion group that belongs to a definite term of the unperturbed atom, regarding it as a representation of the new substitution group, which indeed is a subgroup of the former group; one thus obtains the number and multiplicities of the terms into which the given unperturbed term breaks up when the symmetry is diminished in the way in question. As is well known, for this, it makes no difference whether

[1](E. Wigner, Ztschr. f. Phys. Vol. 43, p. 624. (1927).

the perturbation is small or large, or in general what is its detailed form, or whether one calculates the perturbation energy in first order or in arbitrarily high order: The specification of a definite symmetry of the perturbation potential is entirely sufficient.

To accomplish the reduction of the 21 + 1 dimensional representation of the rotation group, taken as a representation of a definite crystal symmetry group (for example, the octahedral group), we apply the fundamental theorem of group theory: Every reducible representation of a group can be decomposed into its irreducible constituents in one and only one way, and in the reducible representation, the character of each group element is equal to the sum of the characters that belong to the element in the irreducible representations. Accordingly we must know the character of every symmetry operation of the crystal that leaves the nucleus of the atom in its position, both in the irreducible representations of the substitution group of the crystal atom and in the irreducible representations of the substitution group of the free atom.

Every symmetry operation of the crystal atom can now be made up of a pure rotation and possibly an inversion in the nucleus, just as for the symmetry operations of the free atom. But we need only calculate the character of a prescribed rotation in the $2\ell + 1$ dimensional representation of the rotation group; if one adjoins to the rotation an inversion, the character is simply multiplied by +1 or −1, according to whether one is dealing with a positive or a negative term[1]. The most general rotation is that around an arbitrary axis through the angle Φ. Its simplest representation is given by the transformation of the spherical harmonics with lower index l referred to the axis of rotation:

$$f_\mu(x) = P_l^\mu(\cos\vartheta)\, e^{i\mu\varphi}.$$

Our rotation R is to leave ϑ unchanged and carry φ over into $\varphi + \Phi$, and thus takes $f_\mu(x)$ over into

$$f_\mu(Rx) = P_l^\mu(\cos\vartheta)\, e^{i\mu(\varphi+\Phi)},$$

i.e., our rotation is represented by the matrix[1]

[1]Cf. E. Wigner and J. v. Neumann, Ztschr. f. Phys. Vol. 49, p. 73, 91 (1928).

$$\begin{pmatrix} e^{-il\Phi} & 0 & \dots\dots\dots & 0 \\ 0 & & e^{-i(l-1)\Phi} \;.. & 0 \\ \dots & \dots & \dots\dots\dots & \dots \\ 0 & 0 & \dots\dots\dots & e^{il\Phi} \end{pmatrix}$$

with the character

$$\chi(\Phi) = \frac{\sin(l+\frac{1}{2})\Phi}{\sin\frac{1}{2}\Phi}\,. \tag{1}$$

§ 3. Holohedral and Hemihedral Symmetry.

If an atom in the crystal possesses holohedral symmetry, it must be a symmetry center of the whole crystal. Then its symmetry operations can be classified into two categories of equal numbers of elements, the pure rotations on one hand and the rotations with reflection on the other. The pure rotations form an invariant subgroup of the group. From every class of pure rotations, a class of reflections or rotations with reflection arises by multiplication with the inversion in the nucleus.

For, let us assume that C is a class of the invariant subgroup N consisting of the pure rotations, i.e., if X is any element of the invariant subgroup:

$$X^{-1}\,C\,X = C\,.$$

Let J be the inversion, which naturally commutes with every rotation ($J^2 = E$). It then follows that C is also a class in the entire group:

$$(J\,X)^{-1}\,C\,J\,X = C\,.$$

Further more, JC also forms a class of the group

$$X^{-1}\,J\,C\,X = J\,C,$$

q. e. d. (Cf. example in the following section).

Accordingly, the group consists of precisely twice as many elements in precisely twice as many classes as the invariant subgroup in question, and therefore possesses twice as many representations as the latter, namely, first the positive representations, in which the inversion is represented by the unit matrix and the inversion class JC has the same character as the rotation class C, and second the negative representations, in which the inversion is represented by the negative unit matrix and multiplication of a class C by the inversion results in multiplying the character by -1. By comparison with the definition of Wigner and von Neumann[1]

[1] E. Wigner and J. v. Neumann, loc. cit. p. 91.

one sees at once that: In a position of holohedral symmetry, a positive term of the free atom breaks up into nothing but positive crystal terms, a negative term of the free atom into nothing but negative. Beyond this one need bother no further with the inversion, but has only to carry out the reduction of the spherical rotation group taken as representations of the group consisting of the pure rotations of the crystal around the nucleus of the atom, which we shall call the crystal rotation group (tetragonal, hexagonal rotation group).

By omission of half of all the symmetry elements and half of all symmetry classes, there arises from the holohedral group a hemihedral symmetry group, provided that the remaining elements form a group. If the inversion also belongs to the elements of the hemihedral symmetry group, then one can use the same argument as for holohedral symmetry and confine the consideration to the elements of the group consisting of pure rotations. The splitting up of a term of the free atom then in general gives more components than for holohedral symmetry. If the inversion is not an element of the hemihedral symmetry group, then by multiplication by the inversion one can complete the hemihedral group to the holohedral symmetry group, where as above two classes of the holohedral group correspond to each class of the hemihedral group. Accordingly every hemihedral symmetry group that does not contain the inversion corresponds class for class with the invariant subgroup of the holohedral group that consists of the pure rotations, and thus also has precisely the same representations as the latter. (The invariant subgroup itself of course also forms a hemihedral symmetry group, which we can designate as rotational hemihedrism, but besides this other hemihedrisms without an inversion center are also conceivable).

We shall now call a representation of a hemihedral symmetry group positive-equal to a representation of the rotation-hemihedral group if corresponding classes of the two groups have throughout the same characters. We will call two representations of the groups negative-equal if only the rotation classes of the hemihedral symmetry group have exactly the same characters as the corresponding classes of the rotation-hemihedral group, and on the other hand the characters of the reflection classes of the hemihedral symmetry group are equal to the negatives of those of the corresponding rotation classes in the representation of the rotation-hemihedral group used for the comparison.

For a hemihedral symmetry of the position of the atom in the crystal in which it is not an inversion center, a term of the free atom splits up into exactly as many components as for the associated holohedral symmetry. The representations of the terms that arise from a positive term of the free atom in a position of hemihedral symmetry are positive-equal to the irreducible representations of the crystal rotation group (rotational hemihedrism) which one obtains on the reduction of that representation of the spherical rotation group that belongs to the term of the free atom. On the other hand, from a negative term of the free atom there arise, for a position of hemihedral symmetry, terms whose representations are negative-equal to the corresponding representations of the crystal rotation group.

We now proceed to the consideration of particular symmetry groups.

§ 4. Cubic-holohedral Symmetry (e.g., NaCl, Ca in CaF_2).

We assume that all the rotation axes and mirroring planes of the cubic-holohedral symmetry can be chosen to pass through the nucleus of the atom under consideration. The 24 pure rotations of the cubic-holohedral symmetry group, which are identical with the rotations of the octahedral group, fall into five classes:

[Modern Terminology: E, $3C_2$, $6C_4$, $6C_2'$, $8C_3$]

E identity

C_2 rotations around the three 4-fold axes (cube edges) through angle π (3 elements).

C_3 rotations around the 4-fold axes by $\pm \frac{\pi}{2}$ (6 elements).

C_4 rotations around the six 2-fold axes (face diagonals of the cube) by π (6 elements).

C_5 rotations around the four 3-fold axes (diagonals of cube) by $\pm \frac{2\pi}{3}$ (8 elements).

By multiplication with the inversion there arise from these:

J inversion

JC_2 reflections in the planes perpendicular to the 4-fold axes (3 elements).

JC_3 rotation-reflections around the 4-fold axes (rotation by $\pm \frac{\pi}{2}$ + reflection in the plane perpendicular to the axis) (6 elements).

JC_4 reflections in the planes passing through a 4-fold axis and a 2-fold axis perpendicular to it (6 elements).

JC_5 rotation-reflections around the 3-fold axes (6-fold rotation-reflection axes; rotation by $\pm \frac{\pi}{3}$ and reflection in the plane perpendicular to the axis) (8 elements)

The octahedral group consisting of 24 elements accordingly possesses five irreducible representations. The sum of the squares of the dimensionalities of all the representations must be equal to the number of elements in the group; accordingly, we must write 24 as the sum of 5 squares. The only possibility is the decomposition:

$$24 = 3^2 + 3^2 + 2^2 + 1^2 + 1^2 .$$

That is, the octahedral group possesses two three-dimensional, one two-dimensional, and two one-dimensional representations. Every term of the unperturbed atom of higher than three-fold directional degeneracy <u>must</u> therefore split up in a crystal of cubic-holohedral symmetry. In order now to calculate the characters of the representations of the octahedral group, we start with the formula[1]:

$$h_i h_k \chi_i \chi_k = \chi_1 \Sigma c_{ikl} h_l \chi_l .$$

The products of two classes are

$$\begin{aligned} C_2^2 &= 3E + 2C_2 & C_2 C_3 &= C_3 + 2C_4 \\ C_3^2 &= 6E + 2C_2 + 3C_5 & C_2 C_4 &= 2C_3 + C_4 \\ C_4^2 &= 6E + 2C_2 + 3C_5 & C_2 C_5 &= 3C_5 \\ C_5^2 &= 8E + 8C_2 + 4C_5 & C_3 C_4 &= 4C_2 + 3C_5 \\ & & C_3 C_5 &= C_4 C_5 = 4C_3 + 4C_4 \end{aligned}$$

From this we get the following systems of characters.

TABLE 1

	Representation	Class				
		E	$3C_2$	$6C_3$	$6C_4$	$8C_5$
A_1	Γ_1	1	1	1	1	1
A_2	Γ_2	1	1	−1	−1	1
E	Γ_3	2	2	0	0	−1
T_1	Γ_4	3	−1	1	−1	0
T_2	Γ_5	3	−1	−1	1	0

According to the formulas at the end of Sect. 2, in the $2\ell + 1$ dimensional representation of the rotation group, the characters corresponding to the classes of the octahedral group are as follows:

[1] A. Speiser, Theorie der Gruppen, 1 Aufl. p. 163, 2 Aufl. p. 171.

$$\text{Class E} \quad \text{Rotation by} \quad \Phi = 0 \qquad \chi_E = 2l + 1$$

$$\text{,, } C_2 \text{ u. } C_4 \text{ ,, } \quad , \Phi = \pi \quad \chi_2 = \chi_4 = \frac{\sin\left(l + \frac{1}{2}\right)\pi}{\sin\frac{\pi}{2}} = (-1)^l$$

$$\text{,, } C_3 \text{ ,, } \quad \text{,, } \Phi = \frac{\pi}{2} \qquad \chi_3 = \frac{\sin\left(\frac{l}{2} + \frac{1}{4}\right)\pi}{\sin\frac{\pi}{4}} = (-)^{\left[\frac{l}{2}\right]}$$

$$\text{,, } C_5 \text{ ,, } \quad \text{,, } \Phi = \frac{2\pi}{3} \qquad \chi_5 = \frac{\sin\left(\frac{2l}{3} + \frac{1}{3}\right)\pi}{\sin\frac{\pi}{3}}$$

$$= \begin{cases} 1, & \text{for } l = 3m \\ 0, & \text{,, } l = 3m+1 \\ -1, & \text{,, } l = 3m+2. \end{cases}$$

In detail, one has corresponding to the representations of the spherical rotation group the characters for the individual classes of the octahedral group as shown in the following table, and from them one obtains the decompositions into irreducible representations of the octahedral group shown at the right:

TABLE 2

Characters of the classes of the octahedral group in the $2l + 1$ dimensional representation of the rotation group						Decomposition of the $2l + 1$ dimensional representation of the rotation group into irreducible representations of the octahedral group	Number of Terms
l	E	C_2	C_3	C_4	C_5		
0	1	1	1	1	1	Γ_1	1
1	3	−1	1	−1	0	Γ_4	1
2	5	1	−1	1	−1	$\Gamma_3 + \Gamma_5$	2
3	7	−1	−1	−1	1	$\Gamma_2 + \Gamma_4 + \Gamma_5$	3
4	9	1	1	1	0	$\Gamma_1 + \Gamma_3 + \Gamma_4 + \Gamma_5$	4
5	11	−1	1	−1	−1	$\Gamma_3 + 2\Gamma_4 + \Gamma_5$	4
6	13	1	−1	1	1	$\Gamma_1 + \Gamma_2 + \Gamma_3 + \Gamma_4 + 2\Gamma_5$	6
12	25	1	1	1	1	$2\Gamma_1 + \Gamma_2 + 2\Gamma_3 + 3\Gamma_4 + 3\Gamma_5$	11

According to this, S and P terms of the unperturbed atom do not split up when one places the atom in a position of cubic holohedral symmetry in a crystal. From the D terms on, every term splits up, and into at most as many components as the value of the azimuthal quantum number l, accordingly into fewer components than appear in the Stark effect in a homogeneous field[1]. For l = 12 (in general for l = = 12m the characters of all classes of the octahedral group are equal to +1, only

[1] When two terms with the same representation appear in the reduction of the representation, these are of course nevertheless different, just as also their eigenfunctions have different detailed angular dependences and transform in the same way only for the special symmetry operations of the crystal atom. Concerning this, see Sections 11, 18, 22.

the character of the unit element being equal to 25 (or 24m + 1). Accordingly, the 25-dimensional representation of the rotation group, reduced to irreducible parts as a representation of the octahedral group, contains the regular representation exactly once, and in addition one extra occurrence of the identical representation. (Indeed, in the regular representation, all the characters are zero, except that the one of the unit element is equal to the number of elements in the group, here 24). In general it follows that the 2(12m + k) + 1 dimensional representation of the rotation group contains the regular representation of the octahedral group m times, and in addition, the same representations as the 2k + 1 dimensional representation of the rotation group. If one carries out the reduction of the 24m - (2k + 1) dimensional representation, just the components of the 2k + 1 dimensional representation are lacking to make up the regular representation m times.

Note: Almost the same problem as is treated in this section and in Sections 12 and 13 has been solved already by Ehlert[1] from a quite different approach. For him it was a matter of finding the nuclear vibration functions of prescribed Hund symmetry character with respect to interchange of the protons in a molecule of the type CH_4. In this connection one is led to the investigation of the transformation of spherical surface harmonics on interchange of the vertices of a tetrahedron. These interchanges form a cubic-hemihedral group, which includes the classes E, C_2, C_5, JC_3 and JC_4 of the holohedral group. If, with Ehlert, one omits the interchanges that correspond to a mirroring, then there are three possible symmetry characters $\{\overline{1234}\}$, $\{\overline{123}\}4$, and $\{12\}\{34\}$ for an eigenfunction, which correspond to the representations Γ_1 and Γ_2 or Γ_4 and Γ_5 or Γ_3 respectively, of the octahedral group. For example, one finds the number of spherical harmonics of the lth order of symmetry character $\{\overline{1234}\}$ if one adds the numbers of terms with the representation properties Γ_1 and Γ_2 (with respect to rotations of the octahedron) which arise for cubic symmetry from one term of the unperturbed atom with the azimuthal quantum number l. Similarly, the number of functions with the character $\{12\}\{34\}$ is equal to twice the number of terms Γ_3, because two eigen functions belong to each such term, and the number of functions with the symmetry character $\{\overline{123}\}4$ is equal to three times the total number of terms with the representations Γ_4 and Γ_5. In this way one obtains at once the numbers of eigen functions with prescribed symmetry characters as given by Ehlert. In the actual construction of the eigenfunctions, Ehlert's method is indeed probably preferable (cf. § 12, 13).

§ 5. Hexagonal, Tetragonal, and Rhombic (Holohedral) Symmetries.

a) The hexagonal symmetry group contains 12 pure rotations, which fall into 6 classes:

E identity

C_2 rotation by π around the hexagonal axis (1 element).

C_3 rotation by $\pm\frac{2\pi}{3}$ around the hexagonal axis (2 elements).

C_4 rotation by $\pm\frac{\pi}{3}$ around the hexagonal axis (2 elements).

C_5 rotation around one of the three 2-fold crystal axes, by angle π (3 elements).

C_6 rotation by π around one of the three 2-fold axes perpendicular to those above (3 elements).

[2]W. Ehlert, Ztschr. f. Phys. 51, p. 8. 1928.

In addition to these, one would have the 6 reflection or rotation-reflection classes obtained by multiplying by the inversion; we do not, however, need to bother with these (§ 3). The dimensionalities of the 6 representations are found by writing 12 as a sum of 6 squares, which is possible only in the form

$$12 = 2^2 + 2^2 + 1^2 + 1^2 + 1^2 + 1^2$$

There are four one-dimensional and two two-dimensional representations of the hexagonal rotation group; accordingly, even the P terms of a free atom must split up when the atom is inserted into a crystal in a position of hexagonal-holohedral symmetry.

The products of the classes by twos give:

$$\begin{aligned} C_2^2 &= E & C_2 C_3 &= C_4 \\ C_3^2 = C_4^2 &= 2E + C_3 & C_2 C_4 &= C_3 \\ C_5^2 = C_6^2 &= 3E + 3C_3 & C_2 C_5 &= C_6 \\ C_3 C_4 &= 2C_2 + C_4 & C_2 C_6 &= C_5 \\ C_5 C_6 &= 3C_2 + 3C_4 & C_3 C_5 = C_4 C_6 &= 2C_5 \\ & & C_3 C_6 = C_4 C_5 &= 2C_6 \end{aligned}$$

From this one finds for the characters of the irreductible representations of the hexagonal-holohedral group:

TABLE 3.

	E	C_2	C_3	C_4	C_5	C_6
Γ_1	1	1	1	1	1	1
Γ_2	1	1	1	1	−1	−1
Γ_3	1	−1	1	−1	1	−1
Γ_4	1	−1	1	−1	−1	1
Γ_5	2	2	−1	−1	0	0
Γ_6	2	−2	−1	1	0	0

In the 2 l + 1 dimensional representation of the spherical rotation group, the classes of the hexagonal rotation group have the following characters (end of § 2):

$$\chi_E = 2l + 1$$

$$\chi_2 = \chi_5 = \chi_6 = (-)^l$$

$$\chi_3 = \frac{\sin\frac{2l+1}{3}\pi}{\sin\frac{\pi}{3}} = \begin{cases} 1, & \text{for } l = 3m \\ 0, & 3m+1 \\ -1, & 3m+2 \end{cases}$$

$$\chi_4 = \frac{\sin\frac{2l+1}{6}\pi}{\sin\frac{\pi}{6}} = \begin{cases} 1, & \text{for } l \equiv 0 \text{ or } 2 \pmod 6 \\ 2, & \text{,, } l \equiv 1 \quad (\text{,, } 6) \\ -1, & \text{,, } l \equiv 3 \text{ or } 5 \ (\text{,, } 6) \\ -2, & \text{,, } l \equiv 4 \quad (\text{,, } 6) \end{cases}$$

In detail, one finds

TABLE 4

Characters of the classes of the hexagonal group in the 21 + 1 dimensional representation of the spherical rotation group							Decomposition of the 21 + 1 dimensional representation of the rotation group into irreducible representations of the hexagonal rotation group	Number of Terms
l	E	C_2	C_3	C_4	C_5	C_6		
0	1	1	1	1	1	1	Γ_1	1
1	3	-1	0	2	-1	-1	$\Gamma_2+\Gamma_6$	2
2	5	1	-1	1	1	1	$\Gamma_1+\Gamma_5+\Gamma_6$	3
3	7	-1	1	-1	-1	-1	$\Gamma_2+\Gamma_3+\Gamma_4+\Gamma_5+\Gamma_6$	5
4	9	1	0	-2	1	1	$\Gamma_1+\Gamma_3+\Gamma_4+2\Gamma_5+\Gamma_6$	6
5	11	-1	-1	-1	-1	-1	$\Gamma_2+\Gamma_3+\Gamma_4+2\Gamma_5+2\Gamma_6$	7
6	13	1	1	1	1	1	$2\Gamma_1+\Gamma_2+\Gamma_3+\Gamma_4+2\Gamma_5+2\Gamma_6$	9

The 2 (6m + k) + 1 dimensional representation of the rotation group contains the regular representation of the hexagonal group <u>m</u> times, and in addition the same representations of the hexagonal group as are contained in the 2k + 2 dimensional representation of the spherical rotation group. The 2 1 + 1-fold degenerate term of the free atom splits up, as one can easily calculate, in the hexagonal crystal into $\left[\frac{4}{3}\,1\right] + 1$ terms, of which $2\left[\frac{1}{3}\right] + 1$ are single and the other $\left[\frac{2\,1-1}{3}\right] + 1$ are two-fold degenerate.

b) The <u>tetragonal</u> symmetry group contains 8 rotations in 5 classes:

E identity

C_2 rotation by π around the tetragonal axis.

C_3 rotation by $\pm\frac{\pi}{2}$ around the tetragonal axis (2 elements).

C_4 rotation by π around a two-fold axis perpendicular to the tetragonal axis (2 elements).

C_5 rotatation by π around the bisectors of the angles between the above axes (2 elements).

The group possesses four one-dimensional representations and one with two dimensions; the characters are:

TABLE 5
Characters of the Tetragonal Rotation Group

	E	C_2	C_3	C_4	C_5
Γ_1	1	1	1	1	1
Γ_2	1	1	1	− 1	− 1
Γ_3	1	1	− 1	1	− 1
Γ_4	1	1	− 1	− 1	1
Γ_5	2	− 2	0	0	0

The characters of the tetragonal rotations in the $2l + 1$ dimensional representation of the spherical rotation group are (§ 2):

$$\chi_E = 2l + 1$$
$$\chi_2 = \chi_4 = \chi_5 = (-)^l$$
$$\chi_3 = (-)^{\left[\frac{l}{2}\right]}$$

From this, one obtains the decomposition of the $2l + 1$ dimensional representation of the spherical rotation group into irreducible representations of the tetragonal rotation group, which can be generally stated as follows:

TABLE 6

When l = 4λ | $4\lambda + 1$ | $4\lambda + 2$ | $4\lambda + 3$,
the numbers of occurrences of the representations (Γ_i) . . . (n_i) are:

	4λ	$4\lambda + 1$	$4\lambda + 2$	$4\lambda + 3$
Γ_1	$\lambda + 1$	λ	$\lambda + 1$	λ
Γ_2	λ	$\lambda + 1$	λ	$\lambda + 1$
Γ_3	λ	λ	$\lambda + 1$	$\lambda + 1$
Γ_4	λ	λ	$\lambda + 1$	$\lambda + 1$
Γ_5	2λ	$2\lambda + 1$	$2\lambda + 1$	$2\lambda + 2$

One can easily convince oneself that the relations between the characters are satisfied by the values given. In general the $2l + 1$-fold degenerate term of the free atom splits up into $\left[\frac{3}{2} l\right] + 1$ terms, of which $\left[\frac{l+1}{2}\right]$ are two-fold degenerate and $\left[2\,\frac{l}{2}\right] + 1$ are single.

c) Rhombic Symmetry. Each of the 4 rotations (identity and rotation by π around each of the three axes) forms a class by itself. The group possesses four one-dimensional representations, and every term of the free atom splits up completely into 2 l + 1 single terms. The same thing occurs for still lower symmetry.

§ 6. Orientation of a Half-Integral Angular Momentum in the Crystal. Double-Valued Group Representations.

If the Stark effect splitting in the crystal is small in comparison with the separation of the terms of the same multiplet of the free atom, the total angular momentum j of the atom takes up an orientation in the crystal (every component of a multiplet splits up separately) (§ 1, Case 3). If now the atom to be inserted into the crystal possesses even term multiplicities (odd atomic number), the angular momentum j is half-integral, the same as for a single electron, and has a double-valued representation of the space rotation group. Namely, the rotation around an arbitrary axis by the angle Φ has the character

$$\chi(\Phi) = \frac{\sin(j+\frac{1}{2})\Phi}{\sin\frac{1}{2}\Phi} \tag{2}$$

quite in analogy with Eq. (1). But $j + \frac{1}{2}$ is integral, so that

$$\sin(j+\tfrac{1}{2})(\Phi + 2\pi) = \sin(j+\tfrac{1}{2})\Phi,$$

whereas previously we had, for integral l:

$$\sin(l+\tfrac{1}{2})(\Phi + 2\pi) = -\sin(l+\tfrac{1}{2})\Phi$$

Since, however, in the denominator we have

$$\sin\tfrac{1}{2}(\Phi + 2\pi) = -\sin\tfrac{1}{2}\Phi,$$

it follows that for half-integral j:

$$\chi(2\pi \pm \Phi) = -\chi(\Phi).$$

Every character changes its sign when a rotation through 2π is added, i.e., every character is double-valued. Also the character of the identical rotation can be either

$$\chi(0) = 2l + 1$$

or

$$\chi(2\pi) = -(2l + 1)$$

The only character with a unique value is that for a rotation by π around an arbitrary axis,

$$\chi(\pi) = \chi(3\pi) = 0$$

and in general one has

$$\chi(\Phi) = \chi(4\pi - \Phi).$$

Such a double-valued representation of the spherical rotation group can of course contain only double-valued representations of the crystal rotation groups as irreducible components. In order to obtain these double-valued representations, we introduce the fiction that the crystal is not to go over into itself on rotation by 2π around an arbitrary axis, but only on rotation by 4π. We accordingly define a new group element R, the rotation by 2π (around any axis whatever) and supplement the elements of the crystal rotation group by those that arise from them by multiplication by R. We shall designate the group so obtained, which has twice as many elements as the original group, as the crystal double group, and shall inquire about its irreducible representations. This accomplishes our purpose of representing every element of the simple group by two matrices (a two-valued matrix). This procedure corresponds somewhat to the construction of the Riemann surface for the study of multiple-valued functions.

The double group contains more classes than the simple group, but not twice as many. In general, the rotation by Φ around a definite axis does not belong for the double group to the same class as the rotation around the same axis by $2\pi \pm \Phi$; for the characters of the two rotations are indeed different in the reducible representations of the double group[1]. An exception to this is the case $2\pi - \Phi = \Phi = \pi, \chi(\Phi) = 0$. All classes of rotations of the simple crystal group that contain rotations by π accordingly correspond to one class each of the double group, all others to two classes each. We shall find this confirmed in the study of the individual "double groups".

§ 7. The Tetragonal Double Group.

The simple group of the tetragonal rotations can be completely constructed from the two elements

A = rotation by $\pi/2$ around the tetragonal axis,

B = rotation by π around any two-fold axis;

with $A^4 = B^2 = E$ for the simple group. The classes of the simple group and the elements they contain are:

[1] The double-valued representations of the rotation group are indeed reducible representations of the crystal double group.

E contains the element $E = A^4$

C_2 contains the element A^2

C_3 contains the elements A, A^3

C_4 contains the elements B, A^2B

C_5 contains the elements AB, A^3B

In this group, $(AB)^2 = E$.

We expect that in the double group there will be two classes corresponding to each of the classes E and C_3 of the simple group, and one class corresponding to each of the others. We obtain the double group by setting

$$A^4 = B^2 = R, \quad R^2 = E$$

R commutes with all other elements of the group. One obtains the following classes of the double group:

$$
\begin{aligned}
&E \\
&R \\
&C_2 = A^2,\ A^6 = RA^2 \\
&C_3' = A,\ A^7 \qquad A,\ RA^3 \\
&C_3'' = A^3,\ A^5 \qquad A^3,\ RA \\
&C_4 = B,\ A^2B,\ A^4B = RB = B^3,\ A^6B = (A^2B)^3 \\
&C_5 = AB,\ A^3B,\ A^5B,\ A^7B.
\end{aligned}
$$

The 16 elements of the double group accordingly fall into 7 classes, so that to determine the dimensionalities of the irreducible representations we must split 16 up into a sum of 7 squares:

$$16 = 2^2 + 2^2 + 2^2 + 1^2 + 1^2 + 1^2 + 1^2 .$$

From the relations between the classes:

$$
\begin{aligned}
R^2 &= E & C_2^2 &= 2(E + R) \\
RC_2 &= C_2 & C_2C_3' &= C_2C_3'' = C_3' + C_3'' \\
RC_3' &= C_3'' & C_2C_4 &= 2C_4 \\
RC_3'' &= C_3' & C_2C_5 &= 2C_5 \\
RC_4 &= C_4 & C_3'^2 &= C_3''^2 = 2E + C_2 \\
RC_5 &= C_5 & C_3'C_3'' &= 2R + C_2 \\
 & & C_3'C_4 &= C_3''C_4 = 2C_5 \\
 & & C_3'C_5 &= C_3''C_5 = 2C_4 \\
 & & C_4^2 &= C_5^2 = 4E + 4R + 4C_2 \\
 & & C_4C_5 &= 4C_3' + 4C_3''
\end{aligned}
$$

there follows for the characters of the irreducible representations of the tetragonal double group the scheme of Table 7: The first 5 representations are the single-valued, the last two the double-valued representations of the tetragonal rotation group.

TABLE 7
Characters of the Tetragonal Double Group

	E	R	C_2	C_3'	C_3''	C_4	C_5
Γ_1	1	1	1	1	1	1	1
Γ_2	1	1	1	1	1	−1	−1
Γ_3	1	1	1	−1	−1	1	−1
Γ_4	1	1	1	−1	−1	−1	1
Γ_5	2	2	−2	0	0	0	0
Γ_6	2	−2	0	$\sqrt{2}$	$-\sqrt{2}$	0	0
Γ_7	2	−2	0	$-\sqrt{2}$	$\sqrt{2}$	0	0

We still have to reduce to irreducible components the double-valued representations of the spherical rotation group as representations of the tetragonal double group. The characters of the classes of the tetragonal double group in the 2j + 1 dimensional representation of the spherical rotation group are, by the equation (4)

(2) (2) $$\chi = \frac{\sin(j+\frac{1}{2})\Phi}{\sin\frac{1}{2}\Phi} \quad (j+\tfrac{1}{2} = \text{integer}):$$

Angle of rotation	Character
$\Phi=0$	$\chi_E = 2j+1$
$\Phi=2\pi$	$\chi_R = -(2j+1)$
$\Phi=\pm\pi$	$\chi_2 = \chi_4 = \chi_5 = 0$
$\Phi=\pm\frac{\pi}{2}$	$\chi_3' = \dfrac{\sin\left(j+\frac{1}{2}\right)\frac{\pi}{2}}{\sin\frac{\pi}{4}} = \begin{cases} \sqrt{2}, & \text{when } j\equiv\frac{1}{2} \pmod 4 \\ 0, & \text{,, } j\equiv\frac{3}{2},\frac{7}{2} \;(\text{,, } 4) \\ -\sqrt{2}, & \text{,, } j\equiv\frac{5}{2} \;(\text{,, } 4) \end{cases}$
$\Phi=\pm\frac{3\pi}{2}$	$\chi_3'' = -\chi_3'$

The only essential values are χ_E and χ_3'. All other characters can be uniquely derived from these two, or else vanish. In detail one has:

TABLE 8

j	Characters of the classes in the reducible representation, i.e., in the (2j + 1) dimensional representation of the spherical rotation group		Decomposition of the reducible representation into irreducible components
	E	C_3'	
$^1/_2$	2	$\sqrt{2}$	Γ_6
$^3/_2$	4	0	$\Gamma_6 + \Gamma_7$
$^5/_2$	6	$-\sqrt{2}$	$\Gamma_6 + 2\Gamma_7$
$^7/_2$	8	0	$2\Gamma_6 + 2\Gamma_7$
$4\lambda + j'$	$8\lambda + 2j' + 1$	as for j'	$2\lambda(\Gamma_6 + \Gamma_7)$ plus irreducible components for j = j'

§ 8. Double-Valued Representations of the Hexagonal and Cubic Symmetry Groups

a) Of the 6 classes of the simple hexagonal rotation group, three correspond to a rotation by $\pm\pi$, and consequently to one class each of the hexagonal double group; the other three (E, C_3, C_4) contain rotations by 0, $\pm\frac{2\pi}{3}$, $\pm\frac{\pi}{3}$, and therefore correspond to two classes each of the double group. The double group accordingly contains 24 elements in 9 classes. Besides the 6 representations of the simple group there must accordingly exist three further representations of the double group; the sum of the squares of the dimensionalities of these representations must be equal to the number of new elements appearing in addition to those of the simple group, and thus must be 12; thus all three new representations (the double-valued representations of the simple hexagonal rotation group) are two-dimensional.

From the easily obtained relations between the classes, one finds the values shown in the following table for the characters of the individual classes of the hexagonal double group in the three newly added representations, which as representations of the simple group are double-valued (for the single-valued representations Γ_1 to Γ_6 of the simple group, see Table 3).

TABLE 9
Double-Valued Representations of the Hexagonal Rotation Group

	E	R	C_2	C_3'	C_3'	C_4'	C_4''	C_5	C_6
Γ_7	2	−2	0	1	−1	$\sqrt{3}$	$-\sqrt{3}$	0	0
Γ_8	2	−2	0	1	−1	$-\sqrt{3}$	$\sqrt{3}$	0	0
Γ_9	2	−2	0	−2	2	0	0	0	0

In the 2j + 1 dimensional representation of the spherical rotation group the classes of the hexagonal double group have the following characters (Eq. (2)):

$$\chi_E = 2j + 1 \qquad \chi_R = -\chi_E$$

$$\chi_2 = \chi_5 = \chi_6 = 0$$

$$\chi_3' = \frac{\sin (2j+1)\frac{\pi}{3}}{\sin \frac{\pi}{3}} \begin{cases} 1, & \text{for } j \equiv \frac{1}{2} \pmod 3 \\ -1, & \text{,, } j \equiv \frac{3}{2} \text{ (,, } 3) \\ 0, & \text{,, } j \equiv \frac{5}{2} \text{ (,, } 3) \end{cases}$$

$$\chi_4' = \frac{\sin (2j+1)\frac{\pi}{6}}{\sin \frac{\pi}{6}} \begin{cases} \sqrt{3}, & \text{for } j \equiv \frac{1}{2} \text{ or } \frac{3}{2} \pmod 6 \\ 0, & \text{,, } j \equiv \frac{5}{2} \text{ ,, } \frac{11}{2} \text{ (,, } 6) \\ -\sqrt{3}, & \text{,, } j \equiv \frac{7}{2} \text{ ,, } \frac{9}{2} \text{ (,, } 6) \end{cases}$$

$$\chi_3'' = -\chi_3' \qquad \chi_4'' = -\chi_4'.$$

From this, the decompositions of the double-valued representations of the spherical rotation group into irreducible representations of the hexagonal rotation group are found to be as follows:

TABLE 10

j	Decomposition into irreducible components
$^1/_2$	Γ_7
$^3/_2$	$\Gamma_7 + \Gamma_9$
$^5/_2$	$\Gamma_7 + \Gamma_8 + \Gamma_9$
$^7/_2$	$\Gamma_7 + 2\Gamma_8 + \Gamma_9$
$^9/_2$	$\Gamma_7 + 2\Gamma_8 + 2\Gamma_9$
$^{11}/_2$	$2\Gamma_7 + 2\Gamma_8 + 2\Gamma_9$
$6\lambda + j'$	$2\lambda(\Gamma_7 + \Gamma_8 + \Gamma_9)$ + decomposition for j'

b) The double octahedral group contains 48 elements in 8 classes, since the classes E, C_3, C_5 of the simple octahedral group must each correspond to 2 classes of the double group. The octahedral group accordingly possesses three double-valued representations, comprising one four-dimensional and two two-dimensional representations ($48 = 24 + 4^2 + 2^2 + 2^2$, 24 being the sum of the squares of the dimensionalities of the single-valued representations). The characters of the individual classes of the double octahedral group in the double-valued representations of the octahedral group are:

TABLE 11

Tabelle 11

	E	R	C_2	C_3'	C_3''	C_4	C_5'	C_5''
Γ_6	2	-2	0	$\sqrt{2}$	$-\sqrt{2}$	0	1	-1
Γ_7	2	-2	0	$-\sqrt{2}$	$\sqrt{2}$	0	1	-1
Γ_8	4	-4	0	0	0	0	-1	1

In the 2j + 1 dimensional representation of the rotation group, we have:

$$\chi_E = 2j+1 \qquad \chi_R = -\chi_E$$

$$\chi_2 = \chi_4 = 0$$

$$\chi_3' = \frac{\sin(2j+1)\frac{\pi}{4}}{\sin\frac{\pi}{4}} = \begin{cases} \sqrt{2}, & \text{for } j \equiv \frac{1}{2} \pmod 4 \\ 0 & \text{,, } j \equiv \frac{3}{2} \text{ or } \frac{7}{2} \ (\text{,, } 4) \\ -\sqrt{2} & \text{,, } j \equiv \frac{5}{2} \ (\text{,, } 4) \end{cases}$$

$$\chi_5' = \frac{\sin(2j+1)\frac{\pi}{3}}{\sin\frac{\pi}{3}} = \begin{cases} 1 & \text{for } j \equiv \frac{1}{2} \pmod 3 \\ -1 & \text{,, } j \equiv \frac{3}{2} \ (\text{,, } 3) \\ 0 & \text{,, } j \equiv \frac{5}{2} \ (\text{,, } 3) \end{cases}$$

$$\chi_3'' = -\chi_3' \qquad \chi_5' = -\chi_5'$$

TABLE 12
Irreducible Components of the 2j + 1 Dimensional Double-Valued Representation of the Spherical Rotation Group when Reduced as Two-Dimensional Representation of the Octahedral Group

j	Irreducible components
$^1/_2$	Γ_6
$^3/_2$	Γ_8
$^5/_2$	$\Gamma_7 + \Gamma_8$
$^7/_2$	$\Gamma_6 + \Gamma_7 + \Gamma_8$
$^9/_2$	$\Gamma_6 + 2\Gamma_8$
$^{11}/_2$	$\Gamma_6 + \Gamma_7 + 2\Gamma_8$
$6 + j'$	$\Gamma_6 + \Gamma_7 + 2\Gamma_8$ + components for $j = j'$ with with interchange of Γ_6 with Γ_7
$12\lambda + j'$	$2\lambda(\Gamma_6 + \Gamma_7 + 2\Gamma_8)$ + components for $j = j'$

All angular momenta up to $j = {}^3/_2$ are arbitrarily oriented in the cubic crystal; for larger angular momenta, the directional degeneracy is partly destroyed.

c) The rhombic rotation group possesses one double-valued irreducible representation: $\chi_E = 2$, $\chi_R = -2$, and the characters of the rotations by π around the 3 axes are 0. The 2j + 1 dimensional double-valued representation of the spherical rotation group contains this irreducible representation $j + {}^1/_2$ times.

II. The Zeroth Order Eigen Functions in the Crystal.

§ 9. Eigen function of an Atom with Several Electrons.

The eigenfunctions of a free atom with N electrons depend not only on the distances between the electrons and between electrons and nucleus, but also on the orientation of the entire system in space; but because of the spherical symmetry of the problem the 2l + 1 eigenfunctions that belong to a term with the azimuthal quantum number l are determined for all orientations of the system in space when they are given for one orientation. For example, we can, following Wigner[1], assign the values of the functions arbitrarily on the 3N - 3 dimensional

[1] E. Wigner, Ztschr. f. Phys. 43, p. 624, 640 (1927).

"hypersurface" $x_1 = y_1 = x_2 = 0$, as functions of the remaining coordinates of the electrons, and then the eigenfunctions can be found for any other orientation of the atom in space (any coordinates x_1, y_1, x_2) by means of the $2l + 1$ dimensional representation of the rotation group. The assignment of the eigenfunctions on the hypersurface is of course "arbitrary" only to the extent that the behavior of the eigenfunctions under space rotations of the atom does not depend on the special choice of their values on the hypersurface; the actual eigenfunctions must of course be found by a detailed solution of the Schrödinger equation.

In the crystal, the behavior of the eigenfunctions is much less determined by the symmetry alone; they can be arbitrarily chosen not only on a $3N - 3$ dimensional hypersurface, but in a whole region of the $3N$ dimensional space, without coming into conflict with the transformation properties prescribed by the symmetry. If the position of the atomic nucleus in the crystal admits g symmetry operations, then the eigenfunctions of the atom can be arbitrarily chosen for all positions of the first electron inside a suitably chosen sector of solid angle $\frac{4\pi}{g}$, through which no symmetry element passes, and for arbitrary positions of the other electrons. In the other $g - 1$ regions of equal size, they are then determined in virtue of the group-theoretic representation of the symmetry operations of the crystal that is characteristic for the term in question. The exact determination of the eigenfunctions accordingly requires a much more thoroughgoing explicit solution of the Schrödinger equation than for the free atom.

But now, as is well known, it is an essential feature of the perturbation theory - and it is indeed the perturbation of an atom on inclusion in the crystal that we want to study - that it requires first the specification of eigenfunctions of zeroth order; i.e., from the eigenfunctions of the unperturbed (spherically symmetric) problem one must choose precisely those onto which the eigenfunctions of the perturbed problem join continuously. Accordingly one proceeds as if one also knew completely the eigenfunctions of the atom in the crystal, if one prescribes them on Wigner's hypersurface, and uses the representation of the rotation group in the special form as reduced in terms of representations of the crystal group. Thus the symmetry operations of the crystal transform into each other only those eigenfunctions that belong to the same term of the atom in the crystal. But the general representation of the rotation group in the form reduced in terms of a crystal symmetry group would scarcely be simple. Therefore, we restrict ourselves to setting up the angle-dependent eigenfunctions in zeroth approximation for a single electron in the crystal, regarding its coupling with the other electrons of the atom as suppressed. Such eigenfunctions will, however, be adequate in general also for the evaluation of exchange integrals in the crystal[1], when one combines them with, say, the Hartree radial eigenfunctions, which indeed also hold for an uncoupled electron, and assumes that the inclusion of the atom in the crystal has no essential influence on the electron exchange inside an atom.

§ 10. Zeroth Order Eigenfunctions for an Uncoupled Electron in the Crystal

We first suppose that the Schrödinger equation for an electron of the free atom, considered as uncoupled from the other electrons, has been separated in polar coordinates; to the term with the azimuthal quantum number l there thus belong as angle-dependent eigenfunctions, the surface harmonics of lth order. Then we attempt to use linear combinations of these spherical harmonics to get the correct zeroth order eigenfunctions that belong to a definite term of the electron in the crystal, and that therefore transform under symmetry operations of the crystal atom according to a definite irreducible representation of the symmetry group of the crystal. We shall accordingly unite in a "society" all those spherical harmonics that transform among each other under symmetry operations of the crystal, and then by a suitable choice of linear combinations instead of the original spherical harmonics try to make these societies as small as possible - this corresponds to finding the irreducible representations. Finally there must correspond to every k dimensional irreducible component of the $2l + 1$ dimensional representation of the rotation group one irreducible society of k spherical harmonics of lth order. A society found in this way with a definite representation property for symmetry operations then comprises the correct zeroth order eigenfunctions for the crystal term of just this representation property, if no further term of the same representation property arises from the same term of the unperturbed atom, i.e., if the $2l + 1$ dimensional representation of the rotation group contains that particular irreducible representation of the crystal group only once. For the zeroth order eigenfunctions ψ_i are to be determined by the requirement of the general perturbation theory:

$$\int V\psi_i\psi_k\,dx = 0, \quad i \neq k \; (i = -l, \cdots + l;\; k = -l, \cdots, + l) \tag{3}$$

V is the perturbation potential, x stands for all coordinates, and ψ_i and ψ_k are to transform under operations of the group of the crystal atom according to the

[1]The search for such eigenfunctions was the starting point of the present work. Cf. also Ehlert, Zeitschr. f. Phys., loc. cit.

transformations (a_{ij}) and (b_{kl}). By Wigner's well-known argument one finds ($\underline{g}$ is the number of group elements, and R is a substitution of the group, under which V is accordingly unchanged):

$$\left\{\begin{aligned} \int V\,\psi_i(x)\,\psi_k(x)\,dx &= \frac{1}{g}\sum_R \int V(Rx)\,\psi_i(Rx)\,\psi_k(Rx)\,dx \\ &= \frac{1}{g}\sum_{j,l}\sum_R a^R_{ij}\,b^R_{kl}\int V\,\psi_j(x)\,\psi_l(x)\,dx \end{aligned}\right. \tag{4}$$

= 0 for a $\neq$ b, i.e., for terms of different representations.
= $\delta_{ik}\ c_i$, for a = b.

Accordingly, if only one term belongs to a particular representation and if ψ_i is an eigenfunction for this term, then the integral vanishes both for such ψ_k as belong to other terms and for those that belong to the same term; thus the eigenfunctions ψ_i satisfy the requirement (3) of perturbation theory.

If on the other hand several (n) terms belong to an irreducible representation, then in general the integral (3) does not vanish if by ψ_i and ψ_k one understands two eigenfunctions that belong to two different terms of the same representation. The $\underline{n}$ "societies" that belong to the same irreducible representation of the crystal group thus obviously do not in general as yet comprise the correct eigenfunctions of the problem, and these correct eigenfunctions cannot be determined at all by purely group-theoretical methods, but only by taking into account aspects of the particular problem in question. One must, as usual in perturbation theory, form from the nk spherical harmonics of the $\underline{n}$ societies nk new eigenfunctions ψ_i by taking linear combinations:

$$\Psi_i = \sum_{j=1}^{nk} b_{ij}\,\psi_j \qquad (i = 1, \ldots, n\,k)$$

and determine the coefficients b_{ij} by solving the usual determinantal equation of degree nk. This equation gives for the term value $\underline{n}$ k-fold roots, and at the same time, one obtains the $\underline{k}$ zeroth eigenfunctions belonging to each. These eigenfunctions are not, however, linear combinations of spherical harmonics with universal coefficients; the coefficients depend on the detailed form of the electric field of the crystal. (The relative complication of this case, with several terms belonging to the same representation, is closely connected with the general theorem that terms of the same representation do not cross over each other (cf. § 22)). (An example is given in the next section).

§ 11. Eigenfunctions for Tetragonal and Hexagonal Symmetry

In order now actually to arrange the spherical harmonics into "societies" in such a way that under arbitrary symmetry operations of the crystal, the eigenfunctions of a society transform only among themselves, we consider each symmetry as a product of an interchange of the positive and negative directions of each axis and interchanges of the axes of the crystal with each other. Furthermore we write the spherical harmonics in real form, $\sqrt{2}\ P_l^m(\cos\vartheta)\ {\cos \atop \sin}\ m\,\varphi$; then under change of the directions of the axes every spherical harmonic changes at most its sign, and never goes over into another function. Accordingly, all that we have left to deal with is the interchange of different axes, and we see at once that in a rhombic crystal, every spherical harmonic written in real form makes up a society by itself, because interchanges of different axes do not belong to the group. Thus all eigenvalues are single, as we already showed by group-theoretical arguments.

For tetragonal-holohedral symmetry of the position of the atom, the two two-fold axes X and Y are interchangeable. This interchange carries[1] $\cos\varphi$ over into $\pm\sin\varphi$ and vice versa, and also carries $\cos m\varphi$ over into $\pm\sin m\varphi$ for all odd m, while for even m, $\cos m\varphi$ at most changes its sign: For odd m, $\cos m\varphi$ and $\sin m\varphi$ belong to the same society, but for even m every function forms its own society.

Similarly, for the hexagonal-holohedral symmetry (the three two-fold axes perpendicular to the hexagonal axis are interchangeable) $\cos m\varphi$ and $\sin m\varphi$ belong to the same (two-fold) eigenvalue when M is not divisible by 3, and on the other hand they belong to two distinct single eigenvalues when m is a multiple of 3. In both cases, one obtains the number of distinct components known from the group-theoretical argument (§ 5).

But in both cases we can assert that our eigenfunctions are all correct ones only if the reduction of the $2l + 1$ dimensional representation of the spherical rotation group in terms of the irreducible representations of the crystal group does not give any irreducible representation more than once. For tetragonal symmetry, this is the case for $l \leq 2$, for hexagonal symmetry for $l \leq 3$. For $l = 3$, on the other hand, with tetragonal symmetry, two terms belong to the same irreducible representation of the tetragonal group, namely the two-dimensional representation Γ_5. Thus in this case, of the eigenfunctions we have constructed, the only ones that are already the final correct eigenfunctions in zeroth approximation are those that correspond to a one-dimensional representation of the crystal group, i.e., that

[1]We of course identify the tetragonal axis Z with the axis of the spherical harmonics; moreover, we always use spherical harmonics with the normalization $\int (P_l^m)^2 \sin\ d\ = 1$.

form societies of one member. These are P_3^0, $\sqrt{2}\,P_3^2 \cos 2\varphi$ and $\sqrt{2}\,P_3^2 \sin 2\varphi$. The eigenfunctions belonging to the societies P_3^1 and P_3^3 are, on the other hand, not yet the final ones; one still has to form from them 4 linear combinations ψ_i (i = 1,2,3,4) that satisfy the condition

$$\int V \psi_i \psi_k \, d\tau = \delta_{ik} \varepsilon_i$$

of perturbation theory. In general, we write down the expression

$$\psi_i = \psi_{nl}(r)\,(\alpha_{i1} \sqrt{2}\, P_3^1 \cos\varphi + \alpha_{i2} \sqrt{2}\, P_3^1 \sin\varphi + \alpha_{i3} \sqrt{2}\, P_3^3 \cos 3\varphi + \alpha_{i4} \sqrt{2}\, P_3^3 \sin 3\varphi)$$

and get in the well-known way

$$\sum_{j=1}^{4} \alpha_{ij} (\varepsilon_{jk} - \delta_{jk} \varepsilon_i) = 0 \qquad k = 1, 2, 3, 4,$$

where

$$\varepsilon_{jk} = \int V \psi_j \psi_k \, d\tau$$

$$\left\{ \begin{aligned} \varepsilon_{11} &= 2\int V \psi_{nl}^2(r)\,[P_3^1(\cos\vartheta)]^2 \cos^2\varphi \, d\tau \\ &= 2\int V \psi_{nl}^2 (P_3^1)^2 \sin^2\varphi \, d\tau \\ &= \varepsilon_{22} = \int V \psi_{nl}^2 (P_3^1)^2 \, d\tau \\ \varepsilon_{33} &= \int V \psi_{nl}^2 (P_3^3)^2 \, d\tau = \varepsilon_{44} \\ \varepsilon_{13} = \varepsilon_{31} &= 2\int V \psi_{nl}^2 P_3^1 P_3^3 \cos\varphi \cos 3\varphi \, d\tau \\ &= -2\int V \psi_{nl}^2 P_3^1 P_3^3 \sin\varphi \sin 3\varphi \, d\tau \\ &= -\varepsilon_{24} = -\varepsilon_{42}. \end{aligned} \right.$$

The relations come from the fact that V, the crystal potential, remains unchanged under a rotation by $\frac{\pi}{2}$ around the Z axis. Furthermore it always follows from the invariance of V under reflection in the XZ plane (or rotation by π around X) that all integrals vanish that contain products of cos and sin, e.g.,

$$\varepsilon_{14} = 2\int V P_3^1 P_3^3 \psi_{nl}^2 \cos\varphi \sin 3\varphi \, d\tau = 0.$$

Thus to determine the displacements of the terms, we have the determinantal equation

$$\begin{vmatrix} \varepsilon_{11}-\varepsilon & 0 & \varepsilon_{13} & 0 \\ 0 & \varepsilon_{11}-\varepsilon & 0 & -\varepsilon_{13} \\ \varepsilon_{13} & 0 & \varepsilon_{33}-\varepsilon & 0 \\ 0 & -\varepsilon_{13} & 0 & \varepsilon_{33}-\varepsilon \end{vmatrix} = [(\varepsilon_{11}-\varepsilon)(\varepsilon_{33}-\varepsilon)-\varepsilon_{13}{}^2]^2 = 0$$

We obtain two eigenvalues

$$\varepsilon', \varepsilon'' = \frac{\varepsilon_{11}+\varepsilon_{33}}{2} \pm \sqrt{\left(\frac{\varepsilon_{11}-\varepsilon_{33}}{2}\right)^2 + \varepsilon_{13}{}^2}$$

and as eigenfunctions for ϵ' :

$$\psi_1' = \sqrt{1-\delta}\, P_3{}^1 \cos\varphi + \sqrt{1+\delta}\, P_3{}^3 \cos 3\varphi,$$
$$\psi_2' = -\sqrt{1-\delta}\, P_3{}^1 \sin\varphi + \sqrt{1+\delta}\, P_3{}^3 \sin 3\varphi$$

with $\delta = \dfrac{\varepsilon_{33}-\varepsilon_{11}}{\sqrt{(\varepsilon_{33}-\varepsilon_{11})^2 + 4\varepsilon_{13}{}^2}}$,

and similarly as eigenfunctions for ϵ'' :

$$\psi_1'' = \sqrt{1+\delta}\, P_3{}^1 \cos\varphi - \sqrt{1-\delta}\, P_3{}^3 \cos 3\varphi,$$
$$\psi_2'' = \sqrt{1+\delta}\, P_3{}^1 \sin\varphi + \sqrt{1-\delta}\, P_3{}^3 \sin 3\varphi.$$

In § 22, we shall prove that ϵ_{13} actually does not vanish, so that $2\, P_3{}^1 \begin{smallmatrix}\cos\\ \sin\end{smallmatrix} \varphi$ and $2\, P_3{}^3 \begin{smallmatrix}\cos\\ \sin\end{smallmatrix} 3\varphi$, are not yet the correct zeroth order eigenfunctions, and furthermore that ϵ_{13} and so also δ, and accordingly the zeroth order eigenfunctions themselves depend strongly on the special problem, in particular on the "orbit radius" of the electron and the departure of the symmetry from cubic symmetry.

For $l > 3$ and tetragonal symmetry not only the two-dimensional, but also some of the one-dimensional irreducible representations into which $2l + 1$ dimensional representation of the rotation group reduces, have more than one term belonging to them, and it is more and more rarely that one can use the simple functions $2\, P_l{}^m \begin{smallmatrix}\cos\\ \sin\end{smallmatrix} m\varphi$ as the zeroth order eigenfunctions in the crystal. That is, with increasing azimuthal quantum number the splitting up in the tetragonal and also in the hexagonal crystal becomes more and more unlike the ordinary Stark effect

TABLE 13
Eigenfunctions for the Irreducible Representations of the Tetragonal and Hexagonal Rotation Groups[1]

Tetragonal			Hexagonal		
Representation with		Eigenfunctions	Representation with		Eigenfunctions
Even λ	Odd λ		Even λ	Odd λ	
Γ_1	Γ_2	$P_0^{\lambda}, \sqrt{2}\, P_\lambda^{4\mu} \cos 4\mu\varphi$	Γ_1	Γ_2	$P_\lambda^0, \sqrt{2}\, P_\lambda^{6\mu} \cos 6\mu\varphi$
Γ_2	Γ_1	$\sqrt{2}\, P_\lambda^{4\mu} \sin 4\mu\varphi$	Γ_2	Γ_1	$\sqrt{2}\, P_\lambda^{6\mu} \sin 6\mu\varphi$
Γ_3	Γ_4	$\sqrt{2}\, P_\lambda^{4\mu+2} \cos (4\mu+2)\varphi$	Γ_3	Γ_4	$\sqrt{2}\, P_\lambda^{6\mu+3} \cos (6\mu+3)\varphi$
Γ_4	Γ_3	$\sqrt{2}\, P_\lambda^{4\mu+2} \sin (4\mu+2)\varphi$	Γ_4	Γ_3	$\sqrt{2}\, P_\lambda^{6\mu+3} \sin (6\mu+3)\varphi$
Γ_5	Γ_5	$\sqrt{2}\, P_\lambda^{2\mu+1} {\cos \atop \sin} (2\mu+1)\varphi$	Γ_5	Γ_5	$\sqrt{2}\, P_\lambda^{6\mu\pm2} {\cos \atop \sin} (6\mu \pm 2)\varphi$
			Γ_6	Γ_6	$\sqrt{2}\, P_\lambda^{6\mu\pm1} {\cos \atop \sin} (6\mu \pm 1)\varphi$

splitting in a homogeneous field, with which it agrees completely for l = 1 (and for the hexagonal crystal also for l = 2). The four-foldness or six-foldness of the principal axis manifests itself more and more, and causes departures from the relations that hold for an infinite-fold axis, which gives the symmetry relations for the Stark effect in a homogeneous field.

§ 12. Eigenfunctions for Cubic (-holohedral) Symmetry.

For cubic-holohedral symmetry all three axes X, Y, Z are interchangeable. From this it follows at once that the three first-order spherical harmonics, $P_1^0 = \sqrt{\frac{3}{2}} \cos\vartheta$ and $\sqrt{\frac{3}{2}} \sin\vartheta {\cos \atop \sin} \varphi$, are carried over into each other by symmetry operations of the cubic symmetry group: A P term is not split up (cf. § 4), the choice of the axis for the spherical harmonics is thus arbitrary, and the directional degeneracy of the angular momentum remains unchanged.

If we proceed to the D term, we see at once that by interchanges of the three crystal axes X, Y, Z the zonal spherical harmonic $P_2^0 = \sqrt{\frac{5}{2}}\left(\frac{3}{2}\cos^2\vartheta - \frac{1}{2}\right)$ can only go over into a function that contains cos φ (or sin φ) either not at all or to the second degree, that is, besides P_2^0 itself only $\sqrt{2}P_2^2 \cos 2\varphi = \sqrt{\frac{15}{8}} \sin^2\vartheta$ $(2\cos^2\varphi - 1)$. If, for example, we replace the Z axis by the X axis, P_2^0 goes over into

[1] These eigenfunctions are of course still not the "correct" zeroth order eigenfunctions.

$$P_2{}^0(x) = \sqrt{\frac{5}{2}}\left(\frac{3}{2}\sin^3\vartheta\cos^2\varphi - \frac{1}{2}\right)$$
$$= -\frac{1}{2}P_2{}^0 + \frac{\sqrt{3}}{2}\sqrt{2}\,P_2{}^2\cos 2\varphi\,.$$

Accordingly $P_2{}^0$ and $\sqrt{2}\,P_2{}^2 \cos 2\varphi$ belong to the same eigen value, which is two-fold degenerate. Similarly, interchanges of axes can only take

$$\sqrt{2}\,P_2{}^2\sin 2\varphi = \sqrt{\frac{15}{2}}\sin\vartheta\cos\varphi\sin\vartheta\sin\varphi$$

over into such spherical harmonics as contain $\cos\varphi$ or $\sin\varphi$ (or both) only to the first degree, and thus, besides itself, only into

$$\sqrt{2}\,P_2{}^1\cos\varphi \quad \text{and} \quad \sqrt{2}\,P_2{}^1\sin\varphi \quad \left(P_2{}^1 = \sqrt{\frac{15}{4}}\sin\vartheta\cos\vartheta\right)$$

We denote the two terms into which a D term splits up in a crystal with cubic symmetry by D_γ (two-fold term, representation Γ_3) and D_ϵ (three-fold term, representation Γ_5). We present the corresponding eigenfunctions both in written-out form and also in Ehlert's [1] notation. In this notation

$$(\alpha\beta\gamma) = r^{l+1}\frac{\partial^l}{\partial x^\alpha\,\partial y^\beta\,\partial z^\gamma}\left(\frac{1}{r}\right), \quad l = \alpha + \beta + \gamma\,.$$

Term	Rep.	Eigenfunction, ordinary notation	Eigenfunction, Ehlert form
D_γ	Γ_3	$(2\gamma)_1 = \sqrt{\frac{5}{2}}\left(\frac{3}{2}\cos^2\vartheta - \frac{1}{2}\right) = P_2{}^0$	(002)
		$(2\gamma)_2 = \sqrt{2}\,P_2{}^2\cos 2\varphi = \sqrt{\frac{15}{8}}\sin^2\vartheta\cos 2\varphi$	(200)—(020)
D_ε	Γ_5	$(2\varepsilon)_1 = \sqrt{2}\,P_2{}^2\sin 2\varphi = \sqrt{\frac{15}{8}}\sin^2\vartheta\sin 2\varphi$	(110)
		$(2\varepsilon)_2 = \sqrt{2}\,P_2{}^1\cos\varphi = \sqrt{\frac{15}{2}}\sin\vartheta\cos\vartheta\cos\varphi$	(101)
		$(2\varepsilon)_3 = \sqrt{2}\,P_2{}^1\sin\varphi = \sqrt{\frac{15}{2}}\sin\vartheta\cos\vartheta\sin\varphi$	(011)

[1] W. Ehlert, loc. cit.

In like fashion we obtain the eigenfunctions for the results of the splitting up of an F term in the crystal, by assigning to the same term the eigenfunctions that transform into each other on interchange of the axes.

Term	Rep.	Eigenfunction, ordinary notation	Eigenfunction, Ehlert form
F_β	Γ_2	$(3\beta) = \sqrt{2}\, P_3^2 \sin 2\varphi = \frac{\sqrt{210}}{4} \cos\vartheta \sin^2\vartheta \sin 2\varphi$	(111)
		$(3\delta)_1 = P_3^0 = \sqrt{\frac{7}{2}}\left(\frac{5}{2}\cos^3\vartheta - \frac{3}{2}\cos\vartheta\right)$	(003)
F_δ	Γ_4	$(3\delta)_2 = \sqrt{2}\sqrt{\frac{5}{8}}\, P_3^3 \cos 3\varphi - \sqrt{2}\sqrt{\frac{3}{8}}\, P_3^1 \cos\varphi$	(300)
		$(3\delta)_3 = \sqrt{2}\sqrt{\frac{5}{8}}\, P_3^3 \sin 3\varphi + \sqrt{2}\sqrt{\frac{3}{8}}\, P_3^1 \sin\varphi$	(030)
		$(3\varepsilon)_1 = \sqrt{2}\, P_3^2 \cos 2\varphi$	(201)—(021)
F_ε	Γ_5	$(3\varepsilon)_2 = \sqrt{2}\sqrt{\frac{3}{8}}\, P_3^3 \cos 3\varphi + \sqrt{2}\sqrt{\frac{5}{8}}\, P_3^1 \cos\varphi$	(120)—(102)
		$(3\varepsilon)_3 = \sqrt{2}\sqrt{\frac{3}{8}}\, P_3^3 \sin 3\varphi - \sqrt{2}\sqrt{\frac{5}{8}}\, P_3^1 \sin\varphi$	(012)—(210)

Here as is well known,

$$P_3^{\,1} = \sqrt{\frac{21}{2}}\left(\frac{5}{4}\cos^2\vartheta - \frac{1}{4}\right)\sin\vartheta,$$

$$P_3^{\,3} = \frac{\sqrt{70}}{8}\sin^3\vartheta.$$

Similarly for a G-Term:

Term	Rep.	Eigenfunction, ordinary notation	Eigenfunction, Ehlert form
G_α	Γ_1	$(4\alpha) = \sqrt{\frac{7}{12}}\, P_4^0 + \sqrt{2}\sqrt{\frac{5}{12}}\, P_4^4 \cos 4\varphi$	(400) + (040) + (004)
		$(4\gamma)_1 = \sqrt{2}\, P_4^2 \cos 2\varphi$	(400) − (040)
G_γ	Γ_3	$(4\gamma)_2 = \sqrt{\frac{5}{12}}\, P_4^0 - \sqrt{2}\sqrt{\frac{7}{12}}\, P_4^4 \cos 4\varphi$	2·(004) − (400) − (040)
		$(4\delta)_1 = \sqrt{2}\, P_4^4 \sin 4\varphi$	(310) − (130)
G_δ	Γ_4	$(4\delta)_2 = \sqrt{2}\sqrt{\frac{7}{8}}\, P_4^1 \cos\varphi - \sqrt{2}\sqrt{\frac{1}{8}}\, P_4^3 \cos 3\varphi$	(103) − (301)
		$(4\delta)_3 = \sqrt{2}\sqrt{\frac{7}{8}}\, P_4^1 \sin\varphi + \sqrt{2}\sqrt{\frac{1}{8}}\, P_4^3 \sin 3\varphi$	(031) − (013)

Term	Rep	Eigenfunction, ordinary notation	Eigenfunction, Ehlert form
G_ε	Γ_8 Γ_5	$(4\varepsilon)_1 = \sqrt{2}\, P_4^2 \sin 2\varphi$	(112)
		$(4\varepsilon)_2 = \sqrt{2}\sqrt{\frac{1}{8}}\, P_4^1 \cos\varphi + \sqrt{2}\sqrt{\frac{7}{8}}\, P_4^3 \cos 3\varphi$	(121)
		$(4\varepsilon)_3 = \sqrt{2}\sqrt{\frac{1}{8}}\, P_4^1 \sin\varphi - \sqrt{2}\sqrt{\frac{7}{8}}\, P_4^3 \sin 3\varphi$	(211)
		$P_4^0 = \sqrt{\frac{9}{2}}\left(\frac{35}{8}\cos^4\vartheta - \frac{15}{4}\cos^2\vartheta + \frac{3}{8}\right)$	
		$P_4^1 = \frac{3}{8}\sqrt{10}\,(7\cos^3\vartheta - 3\cos\vartheta)\sin\vartheta$	
		$P_4^2 = \frac{3}{8}\sqrt{5}\,(7\cos^2\vartheta - 1)\sin^2\vartheta$	
		$P_4^3 = \frac{3}{8}\sqrt{70}\,\cos\vartheta\sin^3\vartheta$	
		$P_4^4 = \frac{3}{16}\sqrt{35}\,\sin^4\vartheta$	

All our eigenfunctions are the correct zeroth order eigenfunctions for the electron in the crystal atom, for two terms never belong to the same representation[1]. This would first occur in the splitting-up of an H term.

§ 13. Connection between the Splitting-Up of Terms and the Interpretation of Spherical Harmonics as Potentials of Multipoles

Every spherical harmonic of lth order gives the angular dependence of the potential of a multipole of the same order, as can most easily be seen from the Maxwell form of the spherical harmonics,

$$(\alpha\beta\gamma) = r^{l+1}\frac{\partial^l\left(\frac{1}{r}\right)}{\partial x^\alpha\,\partial y^\beta\,\partial z^\gamma}, \quad l = \alpha + \beta + \gamma$$

To every term of the electron for cubic symmetry there now corresponds a definite decomposition ("partition") of $\underline{l}$ into three summands α, β, γ, as was already remarked by Ehlert. Up to $\underline{l} = 3$ the converse of this statement also holds, which makes possible a quite intuitive interpretation of the splitting of terms for cubic symmetry:

For $\underline{l} = 1$ one gets three linearly independent eigenfunctions (100), (010), (001), which correspond to the potentials of dipoles oriented along the X, Y, and

[1]For Ehlert's problem, the construction of the vibrational eigenfunctions of CH_4, a further symmetrization is required, in which one has to form from the eigenfunctions belonging to the same term linear combinations with definite symmetry properties under interchange of the vertices of a tetrahedron.

Z axes. These orientations of a dipole are obviously equally justified, and the three eigenfunctions belong to the same eigenvalue.

For l = 2 we have

$$(200) + (020) + (002) = r^3 \Delta \frac{1}{r} = 0.$$

Accordingly one gets:

a) the two linearly independent eigenfunctions (002) and (200) - (020). These correspond to the potentials of "stretched out" quadrupoles (Fig. 1a), i.e., quadrupoles consisting of oppositely directed dipoles separated along the direction of their axis. Since the directions X, Y, Z are equally justified, both eigenfunctions belong to the same two-fold eigenvalue D_γ.

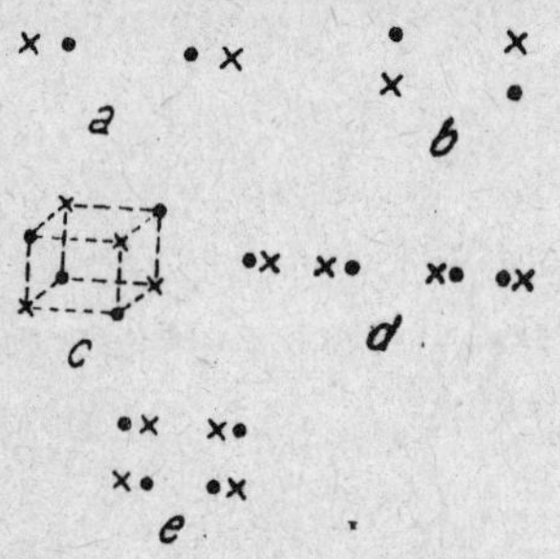

Fig. 1. Quadrupole and octupole classes for cubic symmetry.

b) The three linearly independent eigenfunctions (011), (101), (110) represent the potentials of surface-subtending quadrupoles, which consist of two dipoles separated perpendicularly to their axes (Fig. 1b). The three possible orientations of such a quadrupole in the YZ, ZX, and XY planes, which correspond to the above potentials, are naturally equally justified, and the eigenfunctions belong to the same three-fold eigenvalue D_ϵ.

For l = 3 one gets:

a) The eigenfunction (111), which represents the potential of a volume octupole (Fig. 1c), consisting of two equal and opposite surface-subtending quadrupoles separated perpendicularly to their planes. The eigenfunction belongs to the simple eigenvalue F_β.

b) The eigenfunctions (300), (030), (003), which represent potentials of linearly extended octupoles (Fig. 1d), belong to the three-fold eigenvalue F_δ.

c) The eigenfunctions (120) − (102), (012) − (210), (201) − (021), potential functions of surface-wise extended octupoles[1], belong to the three-fold eigenvalue F_ϵ (Fig. 1e).

For higher multipoles one has more complicated relations; see Ehlert's paper.

III. The Atom Under the Influence of Crystalline Fields of Various Orders of Magnitude.

Hitherto we have dealt in general with the behavior of a quantum-mechanical system possessing a definite angular momentum in an electric field of prescribed symmetry. We must now discuss the three cases listed in the Introduction, which differ in the order of magnitude of the splitting of the terms in the crystal, and also correlate with each other the terms which one obtains in strong, intermediate, and weak crystalline fields.

§ 14. The Angular Distribution of the Electron Density in a Strong Crystalline Field

In the first of the cases distinguished in the Introduction we can in first approximation neglect the exchange interaction of the electrons in the atom in question, and assign a term value and an eigenfunction to the <u>single</u> electron. The actual term value is obtained by addition of the individual electron terms, and the eigenfunction by multiplication of the eigenfunctions of the electrons and "antisymmetrization" of the product[2]. (We speak of "eigenfunction of an individual electron", but are aware that all that can be prescribed is the form of the eigen-function, the quantum cell, and that each electron in the atom has the same probability of being in this quantum cell).

The angular dependence of the eigenfunction of the individual electron is given in the <u>free</u> atom by some spherical harmonic of the lth order or by a linear combination of such functions, if l is the angular momentum of the "orbit" of the electron (cf. § 10). All 2(2l + 1) quantum cells with the same $\underline{l}$ (and the same principal quantum number $\underline{n}$) belong to the same electron term value, and in the absence of the fixing of an axis by an external field it has no meaning to ask to which of the 2(2l + 1) cells an electron belongs: Its probability for being present (density) depends only on the distance from the nucleus and does not prefer any direction in space; its "orbital plane" is arbitrary.

In the crystal, on the other hand, the shell consisting of 2(2l + 1) quantum cells breaks up, according to the irreducible representations of the crystal symmetry

---

[1]The eigenfunctions (210) + (012) = − (030) and cyclic permutations are identical with those given under b).
[2]W. Heitler, Ztschr. f. Phys. Vol. 46, p. 47 (1928).

group, into several subshells, each corresponding to a different electronic term. If the splitting brought about by the crystalline field is sufficiently large or the temperature is sufficiently low, an individual electron will go into the subshell belonging to the lowest term, provided this is not yet "occupied". To every subshell, however, there belongs a quite definite angle-dependent eigenfunction, and the probability for the presence of the individual electron will therefore possess maxima in some directions in the crystal and vanish in others; the angular distribution of the electron density is a definite characteristic of each individual electron term in the crystal.

For example, a p electron in a tetragonal crystal will either go into the shell corresponding to the simple p_β term or into that corresponding to the two-fold term p_ϵ, depending on the special form of the crystalline electric field (cf. § 22). In the former case the angular dependence of its density is given by $\frac{3}{2}\cos^2\vartheta$, so that the density has a maximum along the tetragonal axis ($\vartheta = 0$) and vanishes in the plane of the two-fold axes. In the second case the density is proportional to $\sin^2\vartheta$, and accordingly shows the opposite behavior. If positive ions are located, say, close to the atom in question on the tetragonal axis and at greater distance perpendicular to it, the first case will correspond to a minimum of the energy, and vice versa.

The relationships are most interesting for cubic symmetry of the atom in the crystal. Here a p electron has no preferred orientation at all, all "places" in the p shell are energetically equivalent, and accordingly the probability for the presence of a p electron is in zeroth approximation distributed spherically around the nucleus. On the other hand a d electron can either occupy a place in the d_γ subshell that includes four quantum cells - e.g., a d electron of a negative ion in a NaCl type crystal will always do this - or can go into the six-celled d_ϵ shell, depending on which of the two crystal terms lies deeper. In the first case its eigenfunction will be P_2^0 or with equal probability $\sqrt{2}P_2^2 \cos 2\varphi$, and its density will be given, apart from a factor depending on the distance from the nucleus, by

$$\left.\begin{aligned} \varrho = (P_2^0)^2 + 2\,(P_2^2)^2 \cos^2 2\,\varphi &= \tfrac{5}{4}\,(\tfrac{3}{2}\cos^2\vartheta - \tfrac{1}{2})^2 \\ &+ \tfrac{5}{4}\cdot\tfrac{3}{4}\cdot\sin^4\vartheta\cos^2 2\,\varphi\,. \end{aligned}\right\} \tag{7}$$

(The probabilities obtained from the two eigenfunctions are to be added). This density reaches a maximum $\rho = \frac{5}{4}$ on the three four-fold axes

$$\left(\vartheta = 0, \quad \vartheta = \frac{\pi}{2} \; \varphi = 0, \quad \vartheta = \frac{\pi}{2} \; \varphi = \frac{\pi}{2}\right)$$

and a minimum $\rho = 0$ on the three-fold axes

$$\left(\cos\vartheta = \sqrt{\frac{1}{3}}, \; \vartheta = 54^0\,44', \; \varphi = \frac{\pi}{4}\right).$$

TABLE 14

Density Distribution of a d_γ Electron for Cubic Symmetry

ϑ	$\varphi = 0$	10	20	30	40	45^0
0	1.25	1.25	1.25	1.25	1.25	1.25
10	1.14	1.14	1.14	1.14	1.14	1.14
20	0.87	0.87	0.87	0.86	0.86	0.86
30	0.55	0.54	0.52	0.50	0.49	0.49
40	0.37	0.35	0.30	0.25	0.22	0.21
50	0.34	0.31	0.21	0.10	0.03	0.02
60	0.61	0.55	0.39	0.21	0.10	0.08
70	0.92	0.83	0.61	0.36	0.20	0.18
80	1.16	1.06	0.80	0.50	0.31	0.28
90	1.25	1.14	0.86	0.55	0.34	0.31

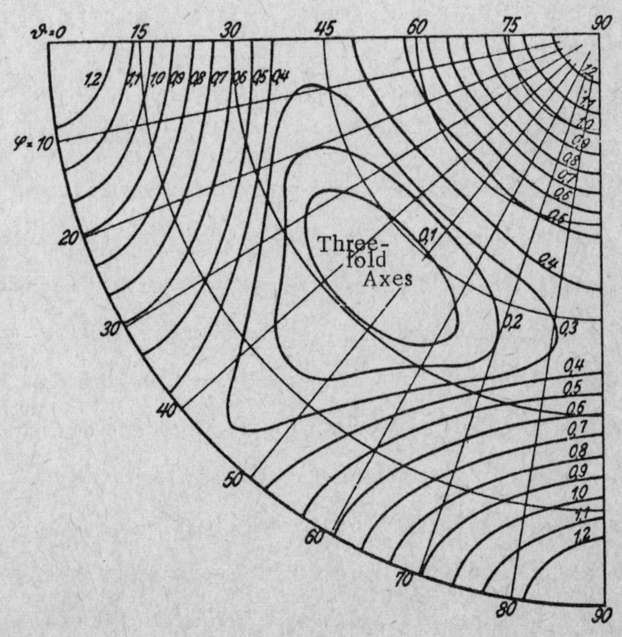

Fig. 2. Electron density distribution of the d_γ electron for cubic symmetry.

Fig. 2 represents the density distribution graphically.

If on the other hand the three-fold term d_ϵ lies lower, then the electron density is greatest on the three-fold axes (diagonals of the cube).

This preference of the electron density for one axis of the crystal is already an effect of zeroth order, just as already in zeroth order the electrons of two hydrogen atoms that unite to form a molecule appear to be attracted to each other[1]. Here, as usual, the zeroth approximation means one that only selects the suitable ones from among the eigenfunctions of the unperturbed problem, without taking account of the perturbation of the eigenfunctions. When further approximations are carried through, of course no density function remains spherically symmetric, and just then the separation of the eigenfunction into product of a radial factor and a spherical harmonic is in general no longer legitimate. In these further approximations, there indeed does not occur any further splitting of the terms, but a further continuous displacement of the term value and deformation of the eigenfunction.

The total charge distribution of a closed shell is of course also no longer spherically symmetric when one calculates the higher approximations for the eigenfunction, i.e., takes into account the deformation by the crystalline field. But in zeroth approximation, the spherical symmetry of closed shells is retained, as also the total term value of a closed shell, in its property as a ^1S term, can never show splitting.

§ 15. Interaction of Electrons of Different Shells as Oriented in the Crystal

We now have the problem of correlating the terms that appear for the atom in the crystal with strong and intermediate values of the crystalline field (cases 1 and 2 of the Introduction). In both cases we begin with the prescription of the quantum numbers of the electrons that are not in closed shells, which we at first regard as not coupled to each other. In the case of the strong crystalline field we then have next to orient the angular momenta of the individual electrons with respect to the field (cf. preceding section), and then to study the interaction of the electrons while preserving their orientation. In the second case (intermediate crystalline field) we have to take into account the interaction in the free atom, i.e., first combine the orbital angular momenta of the electrons into the total orbital angular momentum $\underline{l}$ of the atom and then investigate the orientation of <u>this</u> angular momentum in the crystal. For <u>intermediate</u> crystal splitting, if there is no excitation, the ground state of the atom is at first definitely prescribed, and in the crystal the atom can always be only in a state that arises from the

[1] F. London, Ztschr. f. Phys. Vol. 46, p. 455 (1928).

ground state by Stark effect splitting. In a strong crystalline field, on the other hand, each electron at first seeks for itself the lowest energy state; by interaction of the already oriented electrons an arbitrary term of the atom in the crystal can result, which by no means always needs to go over into the ground term of the free atom on removal of the crystalline field.

It is this state of affairs that we have to investigate group-theoretically. Let us take, say, two electrons with the azimuthal quantum numbers l and λ; then the transformation of their eigenfunctions on rotation of the atom is given by the representations d_l and d_λ of the rotation group. In case 1 (strong crystalline field) we first carry out separately the reductions of d_l and d_λ as representations of the crystal symmetry group (orientation of the individual electron in the crystal); let

$$d_l = \sum \alpha_{lk} \gamma_k ,$$
$$d_\lambda = \sum \alpha_{\lambda\varkappa} \gamma_\varkappa ;$$

The energy of the system composed of the two electrons is then for the present the sum of the energies of the individual electrons, $E_k + E_\varkappa$; the eigenfunction is to be taken as an antisymmetrized product of the eigenfunctions of the individual electrons, and transforms under symmetry operations according to the representation $\gamma_k \cdot \gamma_\varkappa$ of the crystal group. Now we take into account the interaction of the two electrons for fixed orientation, i.e., for fixed terms of the individual electrons. Then the term $E_k + E_\varkappa$ splits up into several terms, which correspond to the irreducible components of $\gamma_k \cdot \gamma_\varkappa$:

$$\gamma_k \cdot \gamma_\varkappa = \sum \beta^i_{k\varkappa} \Gamma_i .$$

Case 2: We first take into account the interaction of the electrons in the free atom, by making the reduction of $d_l \cdot d_\lambda$ as a representation of the rotation group[1]:

$$d_l \cdot d_\lambda = \sum_{|l-\lambda|}^{l+\lambda} D_\varrho .$$

Then we insert the complete atom into the crystal, and have to carry out the reduction of D_ϱ as a representation of the crystal group:

[1] E. Wigner and J. v. Neumann, Ztschr. f. Phys. Vol. 49. p. 73. E. Fues, ibid Vol. 51, p. 817 (1928).

$$D_\varrho = \sum \alpha_{\varrho i} \Gamma_i .$$

The order of succession of the group-theoretical reductions corresponds to the order of succession of the perturbation calculations, i.e., to the relative magnitudes of the contributions of the crystalline field and the electronic interaction to the term value. The number of final terms with a definite representation property must turn out the same by the two procedures. The correlation to each other of the individual terms for strong and intermediate crystalline field can be obtained by the requirement that crystal terms of the entire atom which belong to the same representation of the symmetry group of the crystal must not cross over each other when one thinks of the crystalline field as increasing slowly from intermediate to large strength.

Example: In order to be free from limitations of the Pauli principle, we investigate the terms of a system consisting of two d electrons with different **principal quantum** numbers n_1 and n_2 in a cubic crystal:

Case 1: Large splitting in the crystal.

a) Reduction of the representation of the rotation group belonging to the eigenfunction of the individual electron as a representation of the crystal group (cf. Table 2):

$$d_2 = \gamma_3 + \gamma_5 .$$

Without electronic interaction we obtain four terms: Both electrons can be in the state γ_3, or both in the state γ_5, or one in each state, in which last case, it is not a matter of indifference which is in the state γ_3, since the two electrons are distinguished by their principal quantum numbers. The terms without interaction correspond to the representations $\gamma_3 \cdot \gamma_3$, $\gamma_3 \cdot \gamma_5$, $\gamma_5 \cdot \gamma_3$, and $\gamma_5 \cdot \gamma_5$ of the crystal group.

b) Inclusion of the interaction of the electrons, splitting up of each term in a) in accordance with the irreducible components of its representation[1].

$$\begin{aligned} \gamma_3 \cdot \gamma_3 &= \Gamma_1 + \Gamma_2 + \Gamma_3, \\ \gamma_3 \cdot \gamma_5 &= \Gamma_4 + \Gamma_5, \\ \gamma_5 \cdot \gamma_5 &= \Gamma_1 + \Gamma_3 + \Gamma_4 + \Gamma_5 . \end{aligned}$$

[1]We denote the representation according to which the eigenfunctions of an individual electron transform by small Greek letters, that of the eigenfunction of the entire system by capital Greek letters.

Case 2: Intermediate splitting in the crystal.

a) Interaction of the electrons in the free atom:

$$d_2 \cdot d_2 = D_0 + D_1 + D_2 + D_3 + D_4 .$$

The free atom can be in an S, P, D, F, or G state; which of these is the lowest term can of course not be decided without exact calculation.

b) Orientation of the total orbital angular momentum in the crystal, reduction of that representation of the rotation group which corresponds to the term of the entire atom as representation of the octahedral group (Table 2):

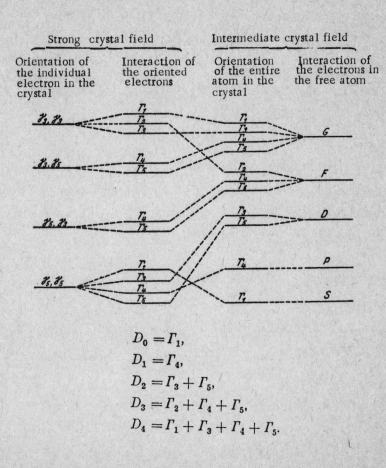

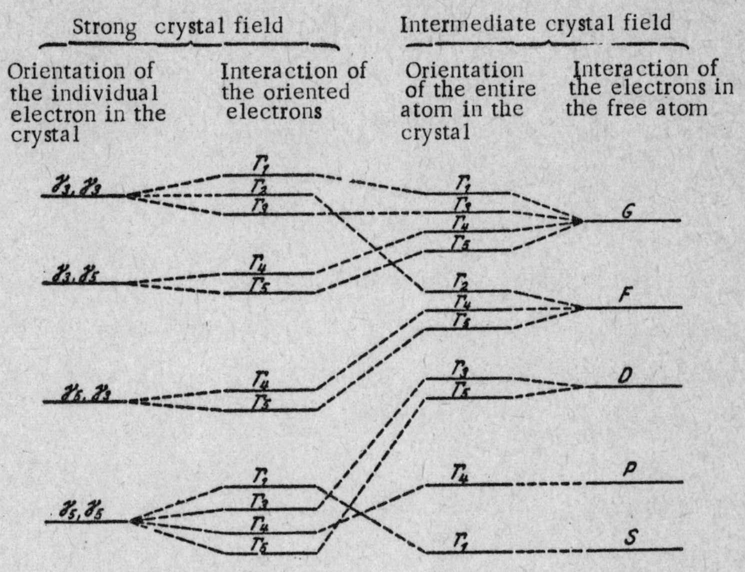

$$D_0 = \Gamma_1,$$
$$D_1 = \Gamma_4,$$
$$D_2 = \Gamma_3 + \Gamma_5,$$
$$D_3 = \Gamma_2 + \Gamma_4 + \Gamma_5,$$
$$D_4 = \Gamma_1 + \Gamma_3 + \Gamma_4 + \Gamma_5.$$

Fig. 3. Terms of a system of two d electrons with different principal quantum numbers, for cubic symmetry.

It is seen that we get, all told, the same terms as before. The correlation is shown in Fig. 3, for arbitrary assumptions about the relative positions of the terms.

§ 16. Interaction of Oriented Electrons of the Same Shell.

In order to obtain only the terms allowed by the Pauli principle for a system of several electrons with the same principal and azimuthal quantum numbers, one proceeds most simply in a way analogous to that used for the corresponding problem in the free atom: One removes the remaining degeneracies wholly or partially, for cubic symmetry say by assuming one axis somewhat extended and thus reducing the symmetry to tetragonal; every degenerate term of a single electron in the cubic system thus splits up into terms of lower degeneracy in the tetragonal system, whose interaction can be treated simply. In this way one obtains the terms of the system of two electrons in a field of tetragonal symmetry, and then has to unite these into terms in the cubic field. The irreducible representations Γ_i of the octahedral group now contain, as one finds on using Tables 1, 5, 7, and 11 to obtain their reductions as representations of the tetragonal rotation group, the following irreducible components[1]:

$$\Gamma_1 = G_1 \quad \Gamma_2 = G_3 \quad \Gamma_3 = G_1 + G_3 \quad \Gamma_4 = G_2 + G_5 \quad \Gamma_5 = G_4 + G_5$$
$$\Gamma_6 = G_6 \quad \Gamma_7 = G_6 \quad \Gamma_8 = G_6 + G_7$$

It is useful to remove also the degeneracy of the two-fold term G_5 of the tetragonal group, by again distinguishing one of the two-fold axes of the tetragonal symmetry group as against the other, and thus reducing the symmetry to the rhombic type. The reduction gives:

$$G_1 = G_3 = \mathfrak{G}_1 \qquad G_2 = G_4 = \mathfrak{G}_2 \qquad G_5 = \mathfrak{G}_3 + \mathfrak{G}_4 .$$

The representations of the rhombic group are all one-dimensional, so that the product of two representations is again an irreducible representation:

$$\mathfrak{G}_1 \mathfrak{G}_i = \mathfrak{G}_i \quad \mathfrak{G}_i^2 = \mathfrak{G}_1 \quad \mathfrak{G}_2 \mathfrak{G}_3 = \mathfrak{G}_4 \quad \mathfrak{G}_3 \mathfrak{G}_4 = \mathfrak{G}_2 \quad \mathfrak{G}_4 \mathfrak{G}_2 = \mathfrak{G}_3 .$$

As an example, we treat the interaction of two d electrons with the same principal quantum number, for cubic symmetry and strong crystalline field. The d shell first splits up into the γ_3 (d_γ) subshell with 4 places (in counting the places we must include the two possible spin directions) and the γ_5 subshell (term d_ϵ with 6 places. If there is one electron in each subshell, then there arise by the interaction all the terms of the two-electron system having representations that are irreducible components of

[1] In this section we write representations of the cubic group with Greek, of the tetragonal group with Latin, and of the rhombic group with German letters.

$$\gamma_3 \gamma_5 = \Gamma_4 + \Gamma_5$$

and indeed always a triplet term and a singlet term, because the spin direction of the two electrons remain arbitrary on account of the different positional eigen-functions. The total quantum weight of these terms if 4.6 = 24.

If, on the other hand, both electrons are in the γ_3 subshell, then out of 16 modes of interaction only $\frac{4 \cdot 3}{2} = 6$ are allowed. The representation properties of the terms of the system of two electrons are again given by the irreducible components of

$$\gamma_3 \cdot \gamma_3 = \Gamma_1 + \Gamma_2 + \Gamma_3,$$

but the multiplicities of the individual terms are still unknown. As described above, we suppose the symmetry decreased to tetragonal; then the γ_3 shell splits up into the g_1 and g_3 subshells of the tetragonal symmetry, with 2 places each. There is now only one way to get both electrons into the g_1 subshell, namely if they have opposite spins. On the other hand, there are four possibilities for having one electron each in g_1 and g_3, because then the spin is arbitrary. We get:

$g_1 \cdot g_1 = G_1$	Quantum weight	1	$m_s = 0$
$g_1 \cdot g_3 = G_3$	,,	4	$m_s = -1\,0\,0\,1$
$g_3 \cdot g_3 = G_1$	,,	1	$m_s = 0$

This means three singlet terms with the representation properties G_1, G_3, G_1 and one singlet term G_3. By comparison with (19) one sees that the term Γ_2 of the two-electron system is a triplet and Γ_1 and Γ_3 are singlet terms.

We treat the interaction of two electrons in the cubic γ_5 shell in the same way:

$$\gamma_5 \cdot \gamma_5 = \Gamma_1 + \Gamma_3 + \Gamma_4 + \Gamma_5 \quad \text{Quantum weight} \quad \frac{6 \cdot 5}{2} = 15.$$

Reduction to tetragonal symmetry:

$$\gamma_5 = g_4 + g_5,$$

$g_4 \cdot g_4 = G_1$	Quantum weight	1	$m_s = 0$,
$g_4 \cdot g_5 = G_5$	,,	8	$m_s = -1\,0\,0\,1$,
$g_5 \cdot g_5 = G_1 + G_2 + G_3 + G_4$	,,	6	$m_s = ?$

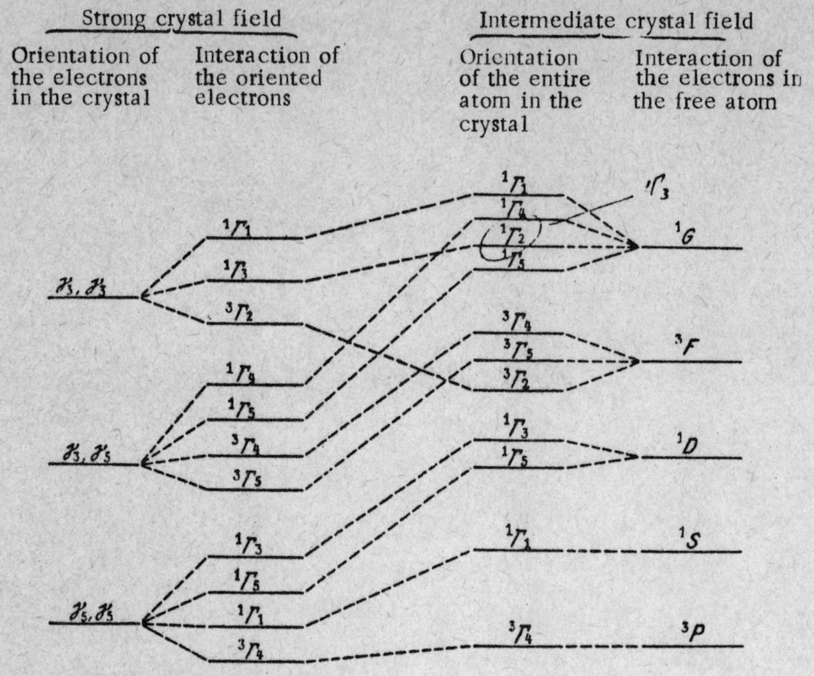

Fig. 4. Terms of a system of two d electrons of the same principal quantum number, for cubic symmetry.

Further reduction to rhombic symmetry:

$$\mathfrak{g}_5 = \mathfrak{g}_3 + \mathfrak{g}_4,$$

$\mathfrak{g}_3 \cdot \mathfrak{g}_3 = \mathfrak{G}_1$	Quantum weight	1	$m_s = 0,$
$\mathfrak{g}_3 \cdot \mathfrak{g}_4 = \mathfrak{G}_2$	"	4	$m_s = -1\,0\,0\,1,$
$\mathfrak{g}_4 \cdot \mathfrak{g}_4 = \mathfrak{G}_1$	"	1	$m_s = 0.$

Combination into two-electron terms for tetragonal symmetry:

$$g_5\, g_5 = {}^1G_1 + {}^1G_2 + {}^1G_3 + {}^3G_4 \quad \text{or} \quad {}^1G_1 + {}^3G_2 + {}^1G_3 + {}^1G_4 \,.$$

Combination into terms for cubic symmetry:

$$\gamma_5\,\gamma_5 = {}^1\Gamma_1 + {}^1\Gamma_3 + {}^1\Gamma_4 + {}^3\Gamma_5 \quad \text{or} \quad {}^1\Gamma_1 + {}^1\Gamma_3 + {}^3\Gamma_4 + {}^1\Gamma_5 \,.$$

Whether Γ_4 or Γ_5 is the triplet term depends on the azimuthal quantum number $\underline{l}$ of the two γ_5 electrons; in our case $\underline{l} = 2$, Γ_5 is the triplet term.

The correlation of the terms of the quantum mechanical system composed of two d electrons can be seen from Fig. 4, for arbitrary assumptions about the positions of the terms.

The interaction of more than two electrons oriented in the crystal can be treated in exactly the same way, and the correlation of the terms in the "strong" and "intermediate" crystalline fields can be completely carried through. We give the representation properties of the crystal terms that can arise from the interaction of several electrons of the same subshell, for cubic symmetry:

TABLE 15

Number of Electrons	Subshell γ_3	Subshell γ_4	Subshell γ_5
1	${}^2\Gamma_3$	${}^2\Gamma_4$	${}^2\Gamma_5$
2	${}^1\Gamma_1 + {}^3\Gamma_2 + {}^1\Gamma_3$	$\left\{ \begin{matrix} {}^1\Gamma_1 + {}^1\Gamma_3 + {}^1\Gamma_4 + {}^3\Gamma_5 \\ {}^1\Gamma_1 + {}^1\Gamma_3 + {}^3\Gamma_4 + {}^1\Gamma_5 \end{matrix} \right.$	$\Big\}$ wie γ_4
3	${}^2\Gamma_3$	$\left\{ \begin{matrix} {}^4\Gamma_1 + {}^2\Gamma_3 + {}^2\Gamma_4 + {}^2\Gamma_5 \\ {}^4\Gamma_2 + {}^2\Gamma_3 + {}^2\Gamma_4 + {}^2\Gamma_5 \end{matrix} \right.$	$\Big\}$ $\begin{matrix} {}^4\Gamma_1 + {}^2\Gamma_3 + {}^2\Gamma_4 + {}^2\Gamma_5 \\ {}^4\Gamma_2 + {}^2\Gamma_3 + {}^2\Gamma_4 + {}^2\Gamma_5 \end{matrix}$
.4	${}^1\Gamma_1$	as for 2 electrons in the shell	
5	—	${}^2\Gamma_4$	${}^2\Gamma_5$
6	—	${}^1\Gamma_1$	${}^1\Gamma_1$

§ 17 Interaction between Orbital Angular Momentum and Spin.

Let an atom be described by its orbital angular momentum $\underline{l}$ and the total spin $\underline{s}$ (term multiplicity 2s + 1). That is, under a common rotation of all electron positions around the nucleus, with spin direction preserved, the wave function of the atom transforms according to the representation D_l of the rotation group; under rotation of the spin direction alone, it transforms according to the representation D_s.[1]

In case 2 of the Introduction (Stark effect splitting in the crystal large in comparison with the multiplet splitting) the orbital angular momentum $\underline{l}$ orients itself independently in the crystal; the entire multiplet is to be taken as the original state and splits up in the crystal into Stark effect components that correspond to the irreducible components Γ_λ of the representation D_l of the spherical rotation group:

$$D_l = \sum \alpha_{l\lambda} \Gamma_\lambda .$$

Each of these Stark effect components now splits up further by the interaction with the spin:

[1]The first roation corresponds to the transformation $P_{\mathfrak{R}}$ of Wigner and von Neumann, loc. cit., the second to the transformation $Q_{\mathfrak{R}}$.

$$\Gamma_\lambda D_s = \sum \alpha_{\lambda\mu} \Gamma_\mu ,$$

This further splitting is of the order of magnitude of the multiplet splitting, but the number of components of a term Γ_λ is in general not equal to 2s + 1.

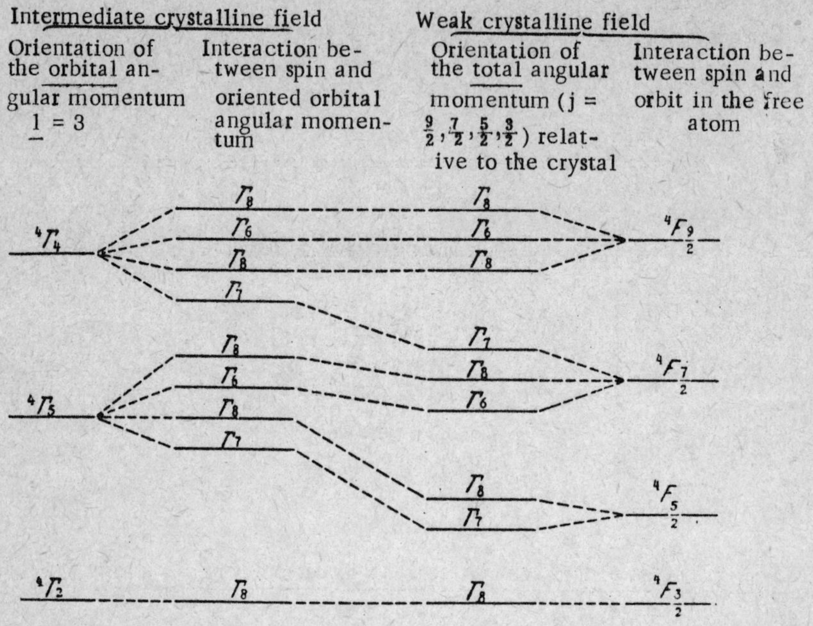

Fig. 5. Orientation of an atom in a ^{4}F state in intermediate and weak fields of cubic symmetry.

In case 3 (crystal splitting small in comparison with multiplet splitting) the spin and orbital angular momentum first combine into the total angular momentum j of the atom:

$$D_l D_s = \sum_{|l-s|}^{l+s} D_j$$

and this orients itself relative to the crystal axes:

$$D_j = \sum \alpha_{j\mu} \Gamma_\mu ,$$

i.e., each multiplet individually undergoes in the crystal a further splitting-up of smaller size than the multiplet splitting. Fig. 5 shows the splitting of a

^4F term for cubic symmetry in intermediate and weak crystalline fields[1].

IV. Magnitude of the Splitting.

§ 18. General Formula for the Term of an Electron in an Ionic Crystal.

After having so far oriented ourselves in a qualitative way about the Stark effect in a crystal, we now wish to calculate also the magnitude of the splitting, at least for the case of the strong crystalline field (every electron with its own term value). It turns out that quite generally the Stark effect in the inhomogeneous field of the crystal is formally an effect of first order, in contrast with the Stark effect in a homogeneous field, for which the first-order perturbation energy always vanishes. Suppose that only one of the terms that arise in the crystalline field from the term n, l of the electron in the free atom belongs to the representation Γ_λ of the crystal group. Then the position of this term λ relative to that of the unperturbed atom is

$$E_\lambda = -\int e V \psi_{nl\lambda}^2 \, d\tau$$

Here V e is the potential energy of the electron in the field of the other atom of the crystal, and $\psi_{nl\lambda}$ is an arbitrary one of the zeroth order eigenfunctions belonging to the term (n, l, λ). $\psi_{nl\lambda}$ can be written as the product of a radial factor and a linear combination of spherical harmonics:

$$\psi_{nl\lambda} = \psi_{nl} \sum_{m=0}^{l}{}' c_{\lambda m} P_l^m(\cos\vartheta) \begin{smallmatrix}\cos\\ \sin\end{smallmatrix} m\varphi .$$

Accordingly we need integrals of the form:

$$K'_{m\mu} = -\int V e\, \psi_{nl}^2(r) P_l^m(\cos\vartheta) P_l^\mu(\cos\vartheta) \begin{smallmatrix}\cos\\ \sin\end{smallmatrix} m\varphi \begin{smallmatrix}\cos\\ \sin\end{smallmatrix} \mu\varphi \, d\tau .$$

By means of such integrals we can evidently also determine all the term values when several terms belong to the same representation of the crystal group (§ 10, 11) and to calculate the eigenvalues one has to set up a special perturbation calculation in which the matrix elements

[1] In practice transitional conditions between intermediate and strong fields (or between intermediate and weak) are probably of most frequent occurrence.

$$\varepsilon_{ii} = -\int V e \psi^2_{nli} d\tau,$$

$$\varepsilon_{ik} = -\int V e \psi_{nli} \psi_{nlk} d\tau$$

are involved.

If there is at least symmetry of the rhombic-holohedral type, all of the integrals $K_{m\mu}$ that contain $\cos m\varphi$ and $\sin \mu\varphi$ mixed vanish. For under reflection in the XZ plane V is unchanged, whereas $\cos m\varphi \sin \mu\varphi$ takes the opposite sign, but the integral $K_{m\mu}$ must be invariant under such changes of the variables of integration, and thus must vanish. We now write:

$$K^{\cos}_{m\mu} = -\int V e \psi^2_{nl}(r) P_l^m(\cos\vartheta) P_l^\mu(\cos\vartheta) \cos m\varphi \cos \mu\varphi \, d\tau$$

$$= -\frac{1}{2}\int V e \psi^2_{nl} P_l^m P_l^\mu [\cos(m-\mu)\varphi + \cos(m+\mu)\varphi],$$

$$K^{\sin}_{m\mu} = -\frac{1}{2}\int V e \psi^2_{nl} P_l^m P_l^\mu [\cos(m-\mu)\varphi - \cos(m+\mu)\varphi].$$

If we rotate the coordinate system by π around the Z axis, V is unchanged, but for odd $m \pm \mu$ $\cos(m \pm \mu)\varphi$ changes its sign, so that:

$$\int V e \psi^2_{nl} P_l^m P_l^\mu \cos(m \pm \mu)\varphi \, d\tau = 0, \quad m \pm \mu \text{ odd.}$$

We now specialize from rhombic to tetrahedral symmetry. Then V also remains unchanged under rotations by $\pi/2$ around Z, while $\cos(m \pm \mu)\varphi$ changes its sign on this rotation if $\frac{m \pm \mu}{2}$ is odd. Accordingly there remain different from zero for tetragonal, and naturally also for cubic, symmetry only the integrals:

$$K_{m\mu} = -\int V e \psi^2_{nl} P_l^m P_l^\mu \cos 4\sigma\varphi \, d\tau \quad (4\sigma = m \pm \mu)$$

and analogously for hexagonal symmetry:

$$\text{(10b)} \quad K_{m\mu} = -\int V e \psi^2_{nl} P_l^m P_l^\mu \cos 6\sigma\varphi \, d\tau \quad (6\sigma = m \pm \mu).$$

We now assume that we have before us an ionic crystal, and write the potential V as the sum of the potentials arising from all ions except the particular one under consideration (perturbing potential):

$$V = \sum_i \frac{l_i}{r_i}.$$

We locate the origin of our system of polar coordinates at the nucleus of the ion under consideration; let the coordinates of the ith ion (source ion) be R_i, Θ, Φ, and those of the argument of V be r, ϑ, φ; the angle between the radii vectors from the nucleus of the ion under consideration to these points is α, and r_i is the distance from the source ion to the argument of V.

We develop the potential produced by the ith ion in spherical harmonics, with the normalization customary in quantum mechanics ($\int (P_l^m)^2 \sin\vartheta \, d\vartheta = 1$):

$$\frac{1}{r_i} = \sum_{s=0}^{\infty} \frac{r^s}{R_i^{s+1}} \sqrt{\frac{2}{2s+1}} P_s^0 (\cos \alpha).$$

By the definition of α we have from the addition theorem of spherical harmonics:

$$\frac{1}{r_i} = \sum_{s=0}^{\infty} \frac{r^s}{R_i^{s+1}} \frac{2}{2s+1} \sum_{\varrho=-s}^{+s} P_s^{\varrho} (\cos \Theta) P_s^{\varrho} (\cos \vartheta) e^{i\varrho(\Phi-\varphi)}$$

and the contribution of the ith ion to the integral $K_{m\mu}$ becomes:

$$K_{m\mu}^{i} = -ee_i \int \psi_{nl}^2(r) r^2 dr \sum_{s=0}^{\infty} \frac{r^s}{R_i^{s+1}} \cdot \frac{2}{2s+1} \sum_{\varrho=-s}^{+s} P_s^{\varrho} (\cos \Theta)$$
$$\cdot \int_0 \sin \vartheta d \vartheta P_s^{\varrho} (\cos \vartheta) P_l^m (\cos \vartheta) P_l^{\mu} (\cos \vartheta)$$
$$\cdot \int_0^{2\pi} d\varphi \cos 4\sigma\varphi (\cos \varrho\varphi \cos \varrho\Phi + \sin \varrho\varphi \sin \varrho\Phi) = -ee_i$$
$$\cdot \int \psi_{nl}^2 r^2 dr \sum_{s=0}^{\infty} \frac{r^s}{R_i^{s+1}} \sqrt{\frac{2}{2s+1}} P_s^{4\sigma} (\cos \Theta) \cos 4\sigma\Phi \cdot \alpha_{lm\mu}^{s\sigma}$$

where

$$\alpha_{lm\mu}^{s\sigma} = \sqrt{\frac{2}{2s+1}} \int P_s^{4\sigma} P_l^m P_l^{\mu} \sin\vartheta d\vartheta$$

is the "expansion coefficient" of the product of spherical harmonics, $P_l^m P_l^\mu$, with respect to the spherical harmonic $P_s^{4\sigma}$. This quantity α can be regarded as a multipole of the sth order of the "transition function" $P_l^m P_l^\mu$; it is a numerical factor of the order of magnitude 1 (cf. § 19).

We now have:

$$K^i_{m\mu} = -e e_i \sum_{s,\sigma} \frac{P_s^{4\sigma}(\cos\Theta)\cos 4\sigma\Phi}{R_i^{s+1}} \overline{r^s}\, \alpha^{s\sigma}_{lm\mu},$$

where

$$\overline{r^s} = \int r^s \psi_{nl}^2(r)\, r^2\, dr \quad .$$

is the mean value of the s power of the distance of the electron from the nucleus of the atom, accordingly a trifle more than the sth power of the "orbit radius" of the electron. The contribution of the moment of sth order to $K^i_{m\mu}$ is accordingly proportional to, say,

$$\frac{e^2}{R_i} \cdot \frac{\overline{r^s}}{R_i^{s+1}} \approx \text{Lattice energy} \cdot \left(\frac{\text{Radius of electron orbit}}{\text{Distance of source ion from ion in question}}\right)^s$$

Since, however, the orbit radius will scarcely be greater than one-fourth of the lattice spacing, the higher moments contribute only very little to $K^i_{m\mu}$, and we can limit ourselves to the consideration of small values of s. But now $s \geq 4\sigma$, so that only $\sigma = 0$ or 1 will be of interest for the calculation of the terms.

A. $\sigma = 0$. According to Eq. (10a) such integrals occur only when $m = \mu$, and thus always have the form $-\int V e \psi_{nl}^2 (P_l^m)^2 \, d\tau$. The consideration of such integrals alone is sufficient, for example, for the calculation of the values of the terms that arise with cubic symmetry from a d or an f term. We at once get:

$$K^i_{mm} = -e e_i \sum_{s=0}^{\infty} \frac{P_s^0(\cos\Theta)}{R_i^{s+1}} \overline{r^s} \cdot \alpha^{s0}_{lmm}.$$

According to the definition of the spherical harmonics one has:

$$\frac{P_s^0(\cos\Theta)}{R_i^{s+1}} = \frac{1}{s!}\left(\frac{\partial^s \frac{1}{R_i}}{\partial z^s}\right)x = y = z = 0,$$

and accordingly when one sums over the contributions of all ions i:

$$K_{mm} = -e\int V\,\psi_{nl}^2(r)[P_l^m(\cos\vartheta)]^2\,d\tau$$
$$= -e\sum_{s=0}^{\infty}\overline{r^s}\cdot\frac{1}{s!}\left(\frac{\partial^s V}{\partial z^s}\right)_0 \alpha_{lm}^s.$$

Here one has:

(12a) $$\alpha_{lm}^s = \alpha_{lmm}^{s0} = \int P_s^0(P_l^m)^2\sin\vartheta\,d\vartheta.$$

For $s = 0$ and $\alpha_{lm}^0 = 1$, $\overline{r^0} = 1$, so that

$$K_{mm} = -eV_0 - e\sum_{s=1}^{\infty}\overline{r^s}\cdot\frac{1}{s!}\left(\frac{\partial^s V}{\partial z^s}\right)_0\cdot\alpha_{lm}^s.$$

$-eV_0$ is the potential energy that the electron would possess if it were concentrated at the nucleus of the ion under consideration. But now the center of gravity of all the terms that arise in the crystal from one term of the free atom is displaced relative to the term of the free atom by precisely the amount

$$E_0 = -\frac{e}{2l+1}\int V\,\psi_{nl}^2(r)\sum_\lambda (P_{l\lambda})^2\,d\tau = -eV_0$$

for the zeroth order density functions $\psi_{nl}^2(r)\,P_{l\lambda}^2$, that belong to the individual crystal terms (n, l) combine according to the addition theorem of the spherical harmonics to form the spherically symmetric electron density $(2l + 1)\psi_{nl}^2(r)$, of the closed shell, and the integration of V over a spherical shell then gives the potential V_0 at the position of the nucleus.

A term of the electron in the crystal atom with the eigenfunction $\psi_{nl}(r)\,P_l^m(\cos\vartheta)\,e^{\pm im\varphi}$ thus has the position, relative to the center of gravity of all crystal terms with the same principal quantum number n and azimuthal quantum number l, given by:

$$(14) \qquad E_m = K_{mm} - E_0 = -e \sum_{s=1}^{\infty} \overline{r^s} \cdot \frac{1}{s!} \left(\frac{\partial^s V}{\partial z^s} \right)_0 \alpha_{lm}^s \cdot$$

As has been said, the main contribution to this expression (14) is made by the first terms of the series. Every term is the product of three independent factors:

1. The mean value of the sth power of the distance of the electron from the nucleus.

2. The sth derivative of the lattice potential V at the position of the nucleus of the ion in question.

3. The intrinsic splitting factor $\alpha^s{}_{lm}$ (multipole moment of the electron density).

B. $\sigma = 1$. Sums of the form (10c) with $\sigma = 1$ occur in the calculation of the following integrals (cf. Eq. (10a)):

$$-e \int V \psi_{nl}^2 [P_l^2 (\cos \vartheta)]^2 (\cos^2 2\varphi - \sin^2 2\varphi) d\tau \qquad (2 + 2 = 4),$$

$$-e \int V \psi_{nl}^2 P_l^1 (\cos \vartheta) P_l^3 (\cos \vartheta) \cos \varphi \cos 3\varphi \, d\tau \qquad (1 + 3 = 4),$$

$$-e \int V \psi_{nl}^2 P_l^{m+2} P_l^{m-2} \cos (m+2)\varphi \cos (m-2)\varphi \, d\tau \qquad ((m+2) - (m-2) = 4).$$

The first type of integral, for example, determines the separation of the two terms with the eigenfunctions $\sqrt{2}\, P_1{}^2 \cos 2\varphi$ and $\sqrt{2}\, P_1{}^2 \sin 2\varphi$ for tetragonal symmetry, and the two other types occur primarily when several terms belong to the same representation (cf. the matrix element ϵ_{13} in § 11, also § 22).

The only important summand in Eq. (10c) is in this case the term with s = 4 (because of the falling off of the terms with increasing s). For this term we have:

$$\sqrt{\frac{2}{9}}\, P_4{}^4 = \sqrt{\frac{35}{128}} \sin^4 \vartheta \qquad .$$

$$\frac{1}{R_i^5} \sqrt{\frac{35}{128}} \sin^4 \Theta \cos 4\Phi = \frac{1}{16} \cdot \frac{1}{4!} \left(4 \frac{\partial^4}{\partial x^4} + 4 \frac{\partial^4}{\partial y^4} - 3 \frac{\partial^4}{\partial z^4} \right) \frac{1}{R_i} \cdot$$

Furthermore, in view of the tetragonal symmetry:

$$\frac{\partial^4 V}{\partial x^4} = \frac{\partial^4 V}{\partial y^4},$$

$$(15)\quad \begin{cases} K^{m\mu} = -e \cdot \frac{1}{4!} \cdot \frac{1}{16}\left(8\,\frac{\partial^4 V}{\partial x^4} - 3\,\frac{\partial^4 V}{\partial z^4}\right)\overline{r^4} \\ \cdot \int \sin^4\vartheta\, P_l^m P_l^\mu \sin\vartheta\, d\vartheta + \text{terms of higher order} \end{cases}$$

For cubic or nearly cubic symmetry we have also:

$$\frac{\partial^4 V}{\partial x^4} = \frac{\partial^4 V}{\partial z^4},$$

$$(15a)\quad \begin{cases} K_{m\mu} = -e \cdot \frac{1}{4!} \cdot \frac{5}{16}\frac{\partial^4 V}{\partial z^4}\overline{r^4} \\ \cdot \int \sin^4\vartheta\, P_l^m P_l^\mu \sin\vartheta\, d\vartheta + \text{terms of higher order} \end{cases}$$

For hexagonal symmetry 6 must be written everywhere instead of 4, and thus the integrals with $\sigma = 0$(cf. A) are at most to be corrected by contributions of multipole moments of 6th order.

§ 19. Expansion of Squares and Products of Spherical Harmonics in Terms of Spherical Harmonics.

We are interested in the expansion coefficients of squares of spherical harmonics in terms of zonal harmonics:

$$(12a)\qquad \alpha_{lm}^s = \sqrt{\frac{2}{2s+1}}\int P_s^0 (P_l^m)^2 \sin\vartheta\, d\vartheta,$$

and also in the following expansion coefficients in terms of the harmonics P_4^4:

$$(16)\quad \begin{cases} \beta_{lm} = \int \sin^4\vartheta (P_l^m)^2 \sin\vartheta\, d\vartheta & \text{for } m = 2 \\ \beta'_{lm} = \int \sin^4\vartheta\, P_l^{m-1} P_l^{m+1} \sin\vartheta\, d\vartheta & \text{for } m = 2 \\ \beta''_{lm} = \int \sin^4\vartheta\, P_l^{m-2} P_l^{m+2} \sin\vartheta\, d\vartheta & \text{for } m \geqslant 2, \end{cases}$$

which occur in Eqs. (14) and (15).

We begin with α^s_{lm}: Since $[P_l^m (\cos\vartheta)]^2$ is a polynomial in $\cos\vartheta$ of degree $2l$, it can be expanded in terms of the first $2l$ zonal harmonics alone:

$$\alpha_{lm}^s = 0 \text{ for } s > 2l.$$

That is, the electron density distribution has no higher multipoles than those of the $2l$ order, and our series (14) for the calculation of the terms in the crystal is always finite. In addition all expansion coefficients with odd index s vanish, because $(P_l^m)^2$ is an even function of $\cos\vartheta$. There remain to be calculated:

$$\alpha_{lm}^{s} \quad \text{for} \quad s = 2\sigma, \quad 0 \leq \sigma \leq l.$$

Here we are primarily interested in the expansion coefficients with small s, which we calculate by the method used by Sommerfeld[1] for the normalization of the spherical harmonics. Accordingly we replace one factor $P_l^m(x)$ by the differential expression:

$$P_l^m(x) = \sqrt{\frac{l-m!}{l+m!} \cdot \frac{2l+1}{2}} \cdot \frac{(1-x^2)^{m/2}}{2^l \cdot l!} \cdot \frac{d^{l+m}}{dx^{l+m}} \cdot (x^2-1)^l$$

and write the other as a polynomial in x, in which we need the two highest powers (the others disappear in differentiations later on)

$$P_l^m(x) = \sqrt{\frac{l-m!}{l+m!} \cdot \frac{2l+1}{2}} \cdot \frac{2l!}{2^l \cdot l!\, l-m!} \cdot (1-x^2)^{m/2} \cdot \left(x^{l-m} - \frac{(l-m)(l-m-1)}{2(2l-1)} x^{l-m-2} + \dots\right)$$

and obtain by partial integrations:

$$\begin{aligned}
\alpha_{lm}^{2} = {} & \frac{2l+1!}{2^{2l+1} l!^2\, l+m!} \int_{-1}^{+1} dx \left(\frac{3}{2}x^2 - \frac{1}{2}\right)(1-x^2)^m \\
& \cdot \left(x^{l-m} - \frac{(l-m)(l-m-1)}{2(2l-1)} x^{l-m-2} \pm \dots\right) \cdot \frac{d^{l+m}}{dx^{l+m}} (x^2-1)^l \\
= {} & (-)^{l+m} \frac{2l+1!}{2^{2l+2} l!^2\, l+m!} \int_{-1}^{+1} (x^2-1)^l \cdot \frac{d^{l+m}}{dx^{l+m}} \Big[(1-x^2)^m \\
& \cdot (3x^2-1)\left(x^{l-m} - \frac{(l-m)(l-m-1)}{2(2l-1)} x^{l-m-2} \pm \dots\right)\Big] \\
= {} & (-)^l \cdot \frac{2l+1!}{2^{2l+2} \cdot l!^2} \int_{-1}^{+1} (x^2-1)^l \cdot \Big[\frac{3}{2}(l+m+2)(l+m+1)x^2 \\
& - \left(1 + 3m + 3\frac{(l-m)(l-m-1)}{2(2l-1)}\right)\Big]\, dx
\end{aligned}$$

[1] A. Sommerfeld, Wellenmechanischer Erganzungsband, p. 63.

$$= (-)^l \cdot \frac{2l+1!}{2^{2l+2}l!^2} \int \Big\{ \frac{3}{2} (l+m+2)(l+m+1)(x^2-1)^{l+1}$$

$$+ \left[\frac{3}{2}(l+m+2)(l+m+1) - \left(1+3m+3\frac{(l-m)(l-m-1)}{2(2l-1)}\right)\right]$$

$$\cdot (x^2-1)^l \Big\} dx$$

$$= -\frac{(l+1)^2}{(2l+2)(2l+3)} \cdot 3 \cdot (l+m+2)(l+m+1)$$

$$+\frac{1}{2}\cdot\left(\frac{3}{2}(l+m+2)(l+m+1) - \left(1+3m+3\frac{(l-m)(l-m-1)}{2(2l-1)}\right)\right).$$

The last form of the expression follows at once from

$$\int_{-1}^{+1} (x^2-1)^\lambda \, dx = (-)^\lambda \cdot \frac{2^{2\lambda} \cdot \lambda!^2}{2\lambda+1!} \cdot 2.$$

By an elementary reduction we obtain finally:

$$\alpha_{lm}^2 = \int (P_l^m)^2 P_2^0 \sqrt{\frac{5}{2}} \sin\vartheta \, d\vartheta = \frac{l(l+1)-3m^2}{(2l+3)(2l-1)}.$$

In precisely similar fashion one calculates α^4_{lm}, requiring however the three highest powers in the polynomial:

$$\alpha_{lm}^4 = \frac{3}{4} \cdot \frac{3l^2(l+1)^2 - 30l(l+1)m^2 + 35m^4 - 6l(l+1) + 25m^2}{(2l+5)(2l+3)(2l-1)(2l-3)}$$

The calculation of the further expansion coefficients now becomes more and more laborious, because more and more terms of the series $P_l^m(x)$ have to be used; these coefficients are indeed also of little importance for our purposes. For small azimuthal quantum numbers (l = 3 and 4) one can nevertheless easily obtain the next two coefficients α^6_{lm} and α^8_{lm}, because the calculation of the coefficients α^{2l-2}_{lm} and α^{2l}_{lm} is again simple. To calculate α^{2l}_{lm} we write the spherical harmonic P^0_{2l} as a differential expression and $(P_l^m)^2$ as a polynomial, in which only the highest power of x is used:

$$(P_l^m)^2 = \binom{2l}{l-m} \frac{2l+1!}{2^{2l+1} \cdot l!^2} (-)^m (x^{2l} \pm \dots)$$

$$\sqrt{\frac{2}{4l+1}} P^0_{2l} = \frac{1}{2^{2l} \cdot 2l!} \frac{d^{2l}}{dx^{2l}} (x^2-1)^{2l}.$$

By partial integration one finds from

$$\alpha_{lm}^{2l} = \binom{2l}{l-m} \frac{(-)^m (2l+1)}{2^{4l+1} \cdot l!^2} \cdot \int (x^{2l} \pm \dots) \frac{d^{2l}}{dx^{2l}} (x^2 - 1)^{2l},$$

The result

$$\alpha_{lm}^{2l} = (-)^m \cdot \binom{2l}{l-m} \cdot \frac{\binom{2l}{l}}{\binom{4l+1}{2l}}.$$

Similarly one obtains:

$$\alpha_{lm}^{2l-2} = (-)^m \cdot \frac{1}{2} \cdot \binom{2l}{l-m} \cdot \frac{\binom{2l-2}{l-1}}{\binom{4l-1}{2l-2}} \cdot \left(1 - \frac{4l-1}{l^2} m^2\right).$$

TABLE 16

s th Order Multipoles of Electron Density Distributed Proportional to the Square of a Spherical Harmonic, $(P_l^m)^2$ = Expansion Coefficient of $(P_l^m)^2$ in Terms of Spherical Harmonics

$$\alpha_{lm}^{s} = \sqrt{\frac{2}{2s+1}} \int_{-1}^{+1} (P_l^m)^2 P_s^0 \, dx.$$

$l =$	1		2			3			
$m =$	0	1	0	1	2	0	1	2	3
$s = 2$	$\frac{2}{5}$	$-\frac{1}{5}$	$\frac{2}{7}$	$\frac{1}{7}$	$-\frac{2}{7}$	$\frac{4}{15}$	$\frac{3}{15} = \frac{1}{5}$	0	$-\frac{5}{15} = -\frac{1}{3}$
$s = 4$	—	—	$\frac{6}{21} = \frac{2}{7}$	$-\frac{4}{21}$	$\frac{1}{21}$	$\frac{6}{33} = \frac{2}{11}$	$\frac{1}{33}$	$-\frac{7}{33}$	$\frac{3}{33} = \frac{1}{11}$
$s = 6$	—	—	—	—	—	$\frac{100}{429}$	$-\frac{75}{429} = -\frac{25}{143}$	$\frac{30}{429} = \frac{10}{143}$	$-\frac{5}{429}$

$l =$	4				
$m =$	0	1	2	3	4
$s = 2$	$\frac{20}{77}$	$\frac{17}{77}$	$\frac{8}{77}$	$-\frac{7}{77} = -\frac{1}{11}$	$-\frac{28}{77} = -\frac{4}{11}$
$s = 4$	$\frac{162}{1001}$	$\frac{81}{1001}$	$-\frac{99}{1001} = -\frac{9}{91}$	$-\frac{189}{1001} = -\frac{27}{143}$	$\frac{126}{1001} = \frac{18}{143}$
$s = 6$	$\frac{20}{143}$	$-\frac{1}{143}$	$-\frac{22}{143} = -\frac{2}{13}$	$\frac{17}{143}$	$-\frac{2}{143}$
$s = 8$	$\frac{490}{2431}$	$-\frac{392}{2431}$	$\frac{196}{2431}$	$-\frac{56}{2431}$	$\frac{7}{2431}$

Expansion Coefficients $\beta = \int P_l^m P_l^\mu \sin^4 \vartheta \sin \vartheta \, d\vartheta$

$l =$	2	3		4		
$m =$	2	2	3	2	3	4
$\mu =$	2	2	1	2	1	0
$\beta =$	$\frac{16}{21}$	$\frac{16}{33}$	$-\frac{16}{165}\sqrt{15}$	$\frac{432}{1001}$	$-\frac{144}{1001}\sqrt{7}$	$\frac{144}{5005}\sqrt{70}$

We now calculate the constants (16). β_{lm} can be reduced to integrals previously treated:

$$\beta_{lm} = \int (P_l^m)^2 \sin^4 \vartheta \sin \vartheta \, d\vartheta$$

$$= \int (P_l^m)^2 \cdot \left(\sqrt{\frac{2}{9}} \cdot \frac{8}{35} \cdot P_4^0 - \sqrt{\frac{2}{5}} \cdot \frac{16}{21} \cdot P_2^0 + \sqrt{2} \cdot \frac{8}{15} \cdot P_0^0 \right).$$

$$(18) \quad \beta_{lm} = 2 \cdot \frac{3l^2(l+1)^2 + 2l(l+1)m^2 + 3m^4 - 14l(l+1) - 15m^2 + 12}{(2l+5)(2l+3)(2l-1)(2l-3)}.$$

For the other two cases one obtains by a calculation like the preceding:

$$(18') \quad \beta'_{lm} = -6 \cdot \sqrt{(l+m+1)(l+m)(l-m+1)(l-m)} \cdot \frac{l(l+1) + m^2 - 4}{(2l+5)(2l+3)(2l-1)(2l-3)}$$

$$(18'') \quad \beta''_{lm} = 6 \cdot \frac{\sqrt{(l+m+2)(l+m+1)(l+m)(l+m-1)(l-m+2)(l-m+1)(l-m)(l-m-1)}}{(2l+5)(2l+3)(2l-1)(2l-3)}.$$

For m = 2 one has from (18) and (18')

$$(18a) \quad \beta_{l2} = 6 \cdot \frac{(l+2)(l+1)l(l-1)}{(2l+5)(2l+3)(2l-1)(2l-3)}.$$

$$(18'a) \quad \beta'_{l2} = -6 \cdot \sqrt{(l+3)(l+2)(l-1)(l-2)} \cdot \frac{l(l+1)}{(2l+5)(2l+3)(2l-1)(2l-3)}.$$

§ 20. The Derivatives of the Lattice Potential.

For cubic symmetry none of the three axes is distinguished, so that from

$$\Delta V = 0$$

one has at once:

$$(19) \quad \frac{\partial^2 V}{\partial x^2} = \frac{\partial^2 V}{\partial y^2} = \frac{\partial^2 V}{\partial z^2} = 0.$$

The crystal field has no effect on the quadrupole moment of the electron distribution; the fourth-order moment is the first that contributes of the term value. On the other hand, for non-cubic symmetry $\frac{\partial^2 V}{\partial z^2}$ is of course finite. Accordingly the term splitting is considerably larger $\left[\text{proportional to} \left(\frac{\text{orbit radius}}{\text{lattice spacing}}\right)^2\right]$ for non-cubic symmetry than for cubic symmetry $\left[\text{proportional to} \left(\frac{\text{orbit radius}}{\text{lattice spacing}}\right)^4\right]$.

We calculate the sth derivative of the lattice potential along one axis direction, $\frac{\partial^s V}{\partial z^s}$, at the location of a positive ion in a crystal of the NaCl type, using Madelung's method[1]. For this purpose, we use the nucleus of our ion as origin of cartesian coordinates. The perturbing potential V(x, 0, 0) consists of a contribution from the row in which the ion itself lies (the X axis), a contribution from the other rows of the plane XY, and a contribution from the other planes z = const. We consider the three contributions to $\frac{1}{s!}\frac{\partial^s V}{\partial x^s}$ individually, writing a for the distance between nearest neighbors and E for the ionic charge.

1. Contribution to the potential from ions of the row in question:

$$V_1(x,0,0) = -\frac{E}{a}\left(\frac{1}{1-\frac{x}{a}} + \frac{1}{3-\frac{x}{a}} + \cdots + \frac{1}{1+\frac{x}{a}} + \frac{1}{3+\frac{x}{a}} + \cdots\right)$$
$$+\frac{E}{a}\left(\frac{1}{2-\frac{x}{a}} + \frac{1}{4-\frac{x}{a}} + \cdots + \frac{1}{2+\frac{x}{a}} + \frac{1}{4+\frac{x}{a}} + \cdots\right).$$

Accordingly:

$$\frac{1}{s!}\left(\frac{\partial^s V_1(x,0,0)}{\partial x^s}\right)_{x=0} = 0$$

When s is odd. For even s one has:

$$\text{(20a)}\quad \begin{cases} \frac{1}{s!}\left(\frac{\partial^s V_1}{\partial x^s}\right)_0 = -\frac{2E}{a^{s+1}}\left(1^{-(s+1)} - 2^{-(s+1)} + 3^{-(s+1)} \pm \cdots\right) \\ \qquad = -\frac{2E}{a^{s+1}} r_{s+1}. \end{cases}$$

The values are $r_3 = 0{,}9016$, $r_5 = 0{,}9722$, $r_7 = 0{,}9926$ etc.

2. Contribution of neighboring rows. The ρth neighboring row at the distance y = ρ a contains a positive or negative ion at the point (0, ρa 0), depending on whether ρ is even or odd, and produces on the x axis the potential [2]

$$V_\varrho(x,0,0) = \frac{4E}{a}(-)^\varrho \sum_{l=1,3,5,\ldots} K_0(\pi l \varrho)\cos\frac{\pi l x}{a}$$

$$\frac{1}{s!}\frac{\partial^s V_\varrho}{\partial x^s} = \frac{4E}{a^{s+1}}(-)^{\varrho+\frac{s}{2}}\frac{\pi^s}{s!}\sum_{l=1,3,5,\ldots} K_0(\pi l \varrho)\, l^s \cos\frac{\pi l x}{a}.$$

for even s. We sum over all rows of the XY plane (each appears twice, to "right" and "left" of the row containing our ion) and set x = 0.

[1] $K_0(x) = H_0^{(1)}(ix)$ = Hankel function of first type with imaginary argument.
[2] E. Madelung, Physikal. Ztschr. Vol. 19, p. 524 (1918).

$$(20b)\quad \frac{1}{s!}\left(\frac{\partial^s V_2}{\partial x^s}\right)_0 = \frac{8E}{a^{s+1}}\cdot(-)^{\frac{s}{2}}\frac{\pi^s}{s!}\sum_{l=1,3,5,\dots} l^s \sum_{\varrho=1,2,3,\dots}(-)^\varrho K_0(\pi l \varrho).$$

3. Contribution of the neighboring plane at distance $z = \rho a$ (at the point $(0, 0, \rho a)$ there is an ion with the charge $(\ )^\rho E$):

$$V'_\varrho = (-)^\varrho \frac{8E}{a}\sum_{l=1,3,5,\dots}\sum_{m=1,3,5,\dots}\frac{e^{-\pi\varrho\sqrt{l^2+m^2}}}{\sqrt{l^2+m^2}}\cos\frac{\pi l x}{a}$$

$$\frac{1}{s!}\frac{\partial^s V'_\varrho}{\partial x^s} = \frac{8E}{a^{s+1}}(-)^{\frac{s}{2}}\frac{\pi^s}{s!}\sum_{l=1,3,5,\dots}\sum_{m=1,3,5,\dots}\frac{e^{-\pi\varrho\sqrt{l^2+m^2}}}{\sqrt{l^2+m^2}}l^s$$

$$(-)^\varrho\cos\frac{\pi l x}{a}.\qquad \text{(für gerade } s\text{)}$$

Contribution of all neighboring planes:

$$(20c)\quad \begin{cases} \frac{1}{s!}\left(\frac{\partial^s V_3}{\partial x^s}\right)_0 = \frac{16E}{a^{s+1}}(-)^{\frac{s}{2}}\frac{\pi^s}{s!} \\ \qquad \sum_{l=1,3,5,\dots}\sum_{m=1,3,5,\dots}(-)^\varrho\frac{e^{-\pi\varrho\sqrt{l^2+m^2}}}{\sqrt{l^2+m^2}}. \end{cases}$$

Thus the total value of the sth derivative of the perturbing potential at the lattice point is:

$$(20)\quad \frac{1}{s!}\left(\frac{\partial^s V}{\partial x^s}\right)_0 = -\frac{2E}{a^{s+1}}r_{s+1} + \frac{8E}{a^{s+1}}(-)^{\frac{s}{2}}\frac{\pi^s}{s!}\sum_{l=1,3,5,\dots} l^s \beta_l,$$

where

$$(20d)\quad \beta_l = \sum_{\varrho=1,2,3,\dots}(-)^\varrho\left(K_0(\pi l\varrho) + 2\sum_{m=1,3,5,\dots}\frac{e^{-\pi\varrho\sqrt{l^2+m^2}}}{\sqrt{l^2+m^2}}\right),$$

$$\beta_1 = -0{,}0450 \qquad \beta_3 = -0{,}650\cdot 10^{-4}$$
$$\beta_5 = -0{,}92\cdot 10^{-7} \qquad \beta_7 = -0{,}138\cdot 10^{-9}.$$

Then we have, for example

$$(20e)\quad \begin{cases} \frac{1}{4!}\left(\frac{\partial^4 V}{\partial z^4}\right)_0 = -3{,}58\frac{E}{a^5} \\ \frac{1}{6!}\left(\frac{\partial^6 V}{\partial z^6}\right)_0 = -0{,}82\frac{E}{a^7}. \end{cases}$$

Now we suppose that our NaCl crystal is stretched somewhat in the direction of one cube edge Z, so that we get a crystal of tetragonal symmetry[1]. Let the ratio of the four-fold axis to the two-fold axes be $\frac{c}{a} = 1 + \epsilon$. Then the second derivatives of the perturbing potential along the directions of the axes of course no longer vanish. The contribution of the ions in the XY plane to $\frac{\partial^s V}{\partial x^s}$ indeed remains constant, but the contribution of the parallel planes $z = \rho\, a\,(1+\epsilon)$ becomes smaller, as is at once obvious physically and comes about formally through the fact that the exponent of every exponential function in Eq. (20c) acquires the factor $\frac{c}{a} = 1 + \epsilon$. Since, now, the parallel planes make a positive contribution[2] to $\frac{\partial^2 V}{\partial x^2}$, we have

$$\frac{\partial^2 V}{\partial x^2} = \frac{\partial^2 V}{\partial y^2} < 0, \qquad \text{for} \qquad \varepsilon > 0$$

and thus because $\Delta V = 0$ accordingly:

$$\frac{\partial^2 V}{\partial z^2} > 0.$$

This means: In a sufficiently stretched NaCl crystal those terms lie lowest for which the second-order moment of the electron density in the direction of the tetragonal axis is positive (cf. Eq. (14)), thus the terms of smallest electric quantum number m (upper index of the spherical harmonic P^m_l). Here the assumption is that the stretching is large enough so that the effect of the fourth-order moment is negligible in comparison with that of the second-order moment (§ 22).. For a shortened crystal of the NaCl type, the situation is of course the opposite, and also the sign is rever sed for the electrons of a negative ion.

§ 21. The Splitting Pattern for Cubic Symmetry.

The magnitude of the term splitting for cubic symmetry is mainly determined by the potential energy of the fourth-order multipole moments of the electron density in the crystal field. According to § 18, the position of the term n, l, γ in relation to the center of gravity of all terms with the same n and l is given by

$$(21) \quad E_\lambda = -e\frac{\overline{r^4}}{4!}\frac{\partial^4 V}{\partial z^4}\int P^2_{l\lambda}\left(P_4^0 + \frac{5}{8}\sin^4\vartheta\cos 4\varphi\right)\sin\vartheta\, d\vartheta\, d\varphi,$$

[1] Such a crystal is not found in nature, but it would be the simplest possible structure of a tetragonal ionic crystal.

[2] For $(-)^{\frac{8}{2}} = -1$, and the main contribution is made by the first parallel plane on each side ($\rho = 1$); on account of the factor $(-)^\rho$, $\frac{\partial^2 V_3}{\partial x^2}$ is thus positive.

where $P_{l\lambda}$ is any one of the angle-dependent eigenfunctions that belong to the term (n, l, λ), accordingly a linear combination of spherical harmonics of lth order.[1] The absolute value of the splitting in a crystal of the NaCl type with monovalent ions is given, according to Eq. (20e), apart from a factor of order of magnitude 1, by

$$D = -e\frac{\overline{r^4}}{4!}\frac{\partial^4 V}{\partial z^4} = \frac{e^2}{a^5}\overline{r^4}\cdot 3{,}58 = \text{say } \frac{e^2}{5a_H}\cdot\frac{3{,}58}{5^4}$$

$$= 1{,}1\cdot 10^{-3}\cdot\frac{e^2}{a_H} = 2\cdot 1{,}1\cdot 10^{-3}\cdot\text{Rydberg const.} = \text{say } 250\,\text{cm}^{-1}.$$

if it is assumed that the radius of the electron orbit is equal to the hydrogen radius a_H and the lattice spacing is $a = 5a_H$. Accordingly the splitting is of the order of magnitude of the multiplet separation. Consequently our calculation of the terms is not exact, since it not only neglects the interaction between spin and orbit, although this is of the same order of magnitude as the effect of the crystalline field, but even the interaction of the individual electrons outside closed shells. Nevertheless we shall carry the calculation further as an illustration of the group-theoretical determination of the splitting.

If we now consider, for example, a positive ion in a crystal of the NaCl type, the relative position of the terms is given, apart from a positive factor D that is a constant for all terms of the same n and l, by:

$$\varepsilon_\lambda = \int P_{l\lambda}^2\left(P_4^0 + \frac{5}{8}\sin^4\vartheta\cos 4\varphi\right)\sin\vartheta\, d\vartheta\, d\varphi .$$

For a d electron, for example, the position of the two-fold degenerate crystal term d_γ relative to the center of gravity of the two terms d_γ and d_ϵ is given by:

$$E_\gamma - E_0 = \frac{2}{7}D,$$

[1]The contribution of the neglected terms of 6th order would be of a size, relative to the terms of 4th order, of say

$$\frac{\frac{\overline{r^6}}{6!}\frac{\partial^6 V}{\partial z^6}}{\frac{\overline{r^4}}{4!}\frac{\partial^4 V}{\partial z^4}} = \frac{0{,}92\frac{E}{a^7}}{3{,}58\frac{E}{a^5}}\,\frac{\overline{r^6}}{\overline{r^4}} = \text{say } \frac{1}{4}\left(\frac{\text{orbit radius}}{\text{lattice spacing}}\right)^2 = \text{say } 0{,}01,$$

i.e., on the basis of data for the NaCl type of crystal, the neglect causes only say 1 percent error.

for the eigenfunction d_γ belongs to P_2^0, and according to Table 16 one has $\int (P_2^0)^2 P_4^0 dx = \frac{2}{7}$. One of the eigenfunctions for the three-fold term d_ϵ is $P_2^1 e^{i\varphi}$, and by Table 16 $\int P_2^1 e^{i\varphi} P_2^1 e^{-i\varphi} \cdot P_4^0 dx = -\frac{4}{21}$, so that[1]

$$E_\varepsilon - E_0 = -\frac{4}{21} D.$$

For a positive ion in a crystal of the NaCl type the two-fold term, for which the maximum electron density is concentrated along the four-fold axes (§ 14) accordingly lies higher than the three-fold term with its electron concentration along the three-fold axes. This is very reasonable: The nearest negative neighbors of the ion lie on the four-fold axes and naturally try to repel the electrons of the positive ion as far from themselves as possible. For a negative ion, the relationships are of course reversed, and they would also reverse on going over to the CsCl type.

An electron in the *f* state can have its maximum probability of occurrence either on the cube diagonals (single term f_β), along the cube edges (three-fold term f_δ), or along the face diagonals (three-fold term f_ϵ). One of the eigenfunctions belonging to the term f_δ is P_3^0, so that

$$E_\delta - E_0 = +\frac{2}{11} D.$$

The term f_β has the eigenfunction $\sqrt{2}\, P_3^2 \sin 2\varphi$, and one of those belonging to f_ϵ is $\sqrt{2}\, P_3^2 \cos 2\varphi$, so that:

$$E_\beta + E_\varepsilon - 2E_0 = 2\int (P_3^2)^2 V \psi_{nl}^2(r)\, d\tau = -2 \cdot \frac{7}{33} D.$$

Making use of the definition of the center of gravity of the terms,

$$7E_0 = E_\beta + 3E_\delta + 3E_\varepsilon$$

one gets for the positions of the terms

[1]As also follows from the definition of the center of gravity of the terms, the relation $2E_\gamma + 3E_\epsilon = 5E_0$.

$$E_\beta - E_0 = -\frac{4}{11} D$$

$$E_\varepsilon - E_0 = -\frac{2}{33} D$$

$$E_\delta - E_0 = +\frac{2}{11} D.$$

For the positive ion of the NaCl crystal the term f_β lies lowest, because here the electron has its density maximum on the cube diagonals, farthest removed from the negative neighbor ions; the term f_δ lies highest (density maximum along the edges); and f_ϵ, with the maximum along the face diagonals, takes an intermediate position. The ratio of the separations of the terms is:

$$E_\delta - E_\varepsilon : E_\varepsilon - E_\beta = 4:5$$

and indeed this holds always for cubic symmetry, not only for the NaCl type of crystal.

Finally we come to the splitting of the g term, and find, using Table 16:

$$E_\alpha - E_0 = \quad 14 \cdot \frac{18}{1001} D$$

$$E_\delta - E_0 = \quad 7 \cdot \frac{18}{1001} D$$

$$E_\gamma - E_0 = \quad 2 \cdot \frac{18}{1001} D$$

$$E_\varepsilon - E_0 = -13 \cdot \frac{18}{1001} D.$$

g_a corresponds to a density maximum on the four-fold axes, g_ϵ on the three-fold axes.

§ 22. The Splitting Pattern for Tetragonal Symmetry. Measure of the "Tetragonality".

The position of a term with the quantum numbers (n, l, γ) relative to the center of gravity of all the terms with the same n and l is:

$$(22) \qquad \begin{cases} E_\lambda - E_0 = -e\,\frac{\overline{r^2}}{2!}\left(\frac{\partial^2 V}{\partial z^2}\right)_0 \alpha_{l\lambda} - e\,\frac{\overline{r^4}}{4!} \\ \cdot\left(\frac{\partial^4 V}{\partial z^4} \cdot \alpha'_{l\lambda} + \frac{1}{8}\left(8\,\frac{\partial^4 V}{\partial x^4} - 3\,\frac{\partial^4 V}{\partial z^4}\right)\beta_{l\lambda}\right), \end{cases}$$

where (cf. § 18):

$$\text{(22a)}\qquad \begin{cases} \alpha_{l\lambda} = \int P_{l\lambda}^{2} P_2^0 \, d\tau \quad \alpha'_{l\lambda} = \int P_{l\lambda}^{2} P_4^0 \, d\tau \\ \beta_{l\lambda} = 2\int P_{l\lambda}^{2} \sin^4 \vartheta \cos 4\varphi \, . \, d\tau \end{cases}$$

Here the assumption is used that the representation γ_λ of the tetragonal group has belonging to it only the one term E_λ with the eigenfunction $P_{1\lambda}$. If one goes over from tetragonal symmetry to cubic symmetry, then $\frac{\partial^2 V}{\partial z^2}$ vanishes, while $\frac{\partial^4 V}{\partial z^4}$ and $\frac{\partial^4 V}{\partial x^4}$ change only relatively little. As long as these fourth-order terms play any part at all, i.e., for nearly cubic symmetry, we can also set $\frac{\partial^4 V}{\partial x^4} = \frac{\partial^4 V}{\partial z^4}$. It now seems suitable to relate the term displacement to the quantity

$$D = -e \frac{\overline{r^4}}{4!} \frac{\partial^4 V}{\partial z^4},$$

in order to make possible a comparison with the relationships for cubic symmetry:

$$\text{(23)}\qquad E_\lambda - E_0 = D\left(\varepsilon_\lambda + \frac{\overline{r^2}}{\overline{r^4}} \cdot \frac{\frac{1}{2!}\left(\frac{\partial^2 V}{\partial z^2}\right)_0}{\frac{1}{4!}\left(\frac{\partial^4 V}{\partial x^4}\right)_0} \alpha_{l\lambda} \right)$$

with

$$\text{(21a)}\quad \varepsilon_\lambda = \alpha'_{l\lambda} + \frac{5}{16}\beta_{l\lambda} = \int P_{l\lambda}^{2}\left(P_4^0 + \frac{5}{8}\sin^4\vartheta \cos 4\varphi\right) d\tau .$$

Here the constant D has only a slight dependence on whether the atom has cubic or tetragonal symmetry. For the NaCl type $D = 3.58 \cdot \frac{e^2}{a^5}$ we are primarily interested in the second quantity in the parentheses, which vanishes for cubic symmetry, and for tetragonal symmetry represents the difference from the cubic case. We call it the effective tetragonality,

$$u = \frac{\overline{r^2}}{\overline{r^4}} \cdot \frac{\frac{1}{2!}\left(\frac{\partial^2 V}{\partial z^2}\right)_0}{\frac{1}{4!}\left(\frac{\partial^4 V}{\partial z^4}\right)_0}$$

and thus have:

$$E_\lambda - E_0 = D(\varepsilon_\lambda + u\,\alpha_{l\lambda}).$$

The effective "tetragonaltiy" determines the relative position of the Stark effect components of an electron term for tetragonal symmetry. It is inversely proportional to the square of the "orbit radius" of the electron. In order to obtain a constant for the symmetry of the atom, we define the absolute tetragonality

$$U = \frac{\frac{1}{2!}\left(\frac{\partial^2 V}{\partial z^2}\right)_0}{\frac{1}{4!}\left(\frac{\partial^4 V}{\partial z^4}\right)_0}$$

as the ratio of the second to the fourth derivative of the perturbing lattice potential with respect to the direction of the tetragonal axis, at the position of the nucleus of the atom. For cubic symmetry U = 0, and we shall call a symmetry nearly cubic when the tetragonality U is small (say smaller than 0.1): on the other hand, for large U we have to do with decidedly tetragonal symmetry[1]. The introduction of the tetragonality is justified by the fact that it is a criterion for the relative positions of the terms, and thus also for the most stable electron distribution in an atom in a prescribed position in the crystal.

To be sure, we have confirmed this significance of the tetragonality so far only for the case in which only one term belongs to each irreducible representation of the tetragonal group. We shall base our proof that it has the same significance (on the same hypothesis $\frac{\partial^4 V}{\partial x^4} \approx \frac{\partial^4 V}{\partial z^4}$); also in the case of several terms with the same representation on the example already begun in § 11 (this can easily be generalized). The two terms that belong to the two-dimensional representation for the azimuthal quantum number l = 3 are given by:

$$E_5', E_5'' = \frac{\varepsilon_{11} + \varepsilon_{33}}{2} \pm \sqrt{\left(\frac{\varepsilon_{11} - \varepsilon_{33}}{2}\right)^2 + \varepsilon_{13}^2}\,.$$

[1] For a stretched-out crystal of the NaCl type, the tetragonality is always negative (cf. § 20), and for small stretchings

$$U \approx -1.71\epsilon\,,$$

and corresponding to this, for an orbit radius of say, 1/4 of the lattice spacing, the effective tetragonality is

$$u \approx -30\,\epsilon = -30\,\frac{c-a}{a}$$

For a shortened crystal axis, the sign is reversed; on the other hand it is independent of the sign of the change of the ion considered.

Here, according to Eq. (5), and by the use of Eqs. (14) and (15) and Table 16, we have

$$\varepsilon_{11} = -eV_0 - \frac{1}{5} e \frac{\overline{r^2}}{2!} \left(\frac{\partial^2 V}{\partial z^2}\right)_0 - \frac{1}{33} e \frac{\overline{r^4}}{4!} \left(\frac{\partial^4 V}{\partial z^4}\right)_0,$$

$$\varepsilon_{33} = -eV_0 + \frac{1}{3} e \frac{\overline{r^2}}{2!} \left(\frac{\partial^2 V}{\partial z^2}\right)_0 - \frac{1}{11} e \frac{\overline{r^4}}{4!} \left(\frac{\partial^4 V}{\partial z^4}\right)_0,$$

$$\varepsilon_{13} = \frac{16}{165}\sqrt{15}\, e \frac{\overline{r^4}}{4!} \cdot \frac{1}{16}\left(8 \frac{\partial^4 V}{\partial x^4} - 3 \frac{\partial^4 V}{\partial z^4}\right)_0 \approx \frac{\sqrt{15}}{33} \frac{\overline{r^4}}{4!} \left(\frac{\partial^4 V}{\partial z^4}\right)_0 .$$

Then

$$\frac{\varepsilon_{11} + \varepsilon_{33}}{2} + eV_0 = D\left(\frac{2}{33} - \frac{u}{15}\right),$$

$$\frac{\varepsilon_{11} - \varepsilon_{33}}{2} = D\left(-\frac{1}{33} + \frac{4u}{15}\right),$$

$$\varepsilon_{13} = -D \cdot \frac{\sqrt{15}}{33} .$$

$$(25) \quad \begin{cases} E_5' - E_0 = D\left(\dfrac{2}{33} - \dfrac{u}{15} + \dfrac{4}{3}\sqrt{\dfrac{u^2}{5^2} - \dfrac{u}{110} + \dfrac{1}{11^2}}\right) \\ E_5'' - E_0 = D\left(\dfrac{2}{33} - \dfrac{u}{15} - \dfrac{4}{3}\sqrt{\dfrac{u^2}{5^2} - \dfrac{u}{110} + \dfrac{1}{11^2}}\right). \end{cases}$$

Here also the position of the terms turns out to depend mainly on the effective tetragonality u; the proportionality factor D is the same as in Eq. (23a).

For small effective tetragonality (almost cubic symmetry) one has by Eq. (25):

$$(25a) \quad \begin{cases} E_5' - E_0 = D\left(\dfrac{2}{11} - \dfrac{2u}{15} + \dfrac{11}{40} u^2 + \cdots\right) \\ E_5'' - E_0 = D\left(-\dfrac{2}{33} \qquad - \dfrac{11}{40} u^2 + \cdots\right). \end{cases}$$

That is: For cubic symmetry (u = 0), E'_5 goes over into the term f_δ, and E''_5 into the term f_ϵ. For a slight departure from cubic symmetry, the term value varies continuously with the departure. For pronounced tetragonal symmetry ($u \gg 1$), on the other hand:

$$E_5' - E_0 = D\left(-\frac{u}{15} + \frac{2}{33} + \frac{4}{15}|u| - \frac{1}{33}\frac{u}{|u|} + \frac{1}{u}(\cdots)\right) .$$

(The positive sign of the square root belongs to E_5', so that u must be inclosed in absolute value marks). For large positive u one gets:

$$(25b) \qquad E_5' - E_0 = D\left(\frac{u}{5} + \frac{1}{33} + \frac{1}{u} \cdots \right),$$

and on the other hand for u < 0:

$$(25c) \qquad E_5' - E_0 = D\left(-\frac{u}{3} + \frac{1}{11} + \frac{1}{u} \cdots \right).$$

Similarly:

$$(25d) \quad \begin{cases} E_5' - E_0 = D\left(-\frac{u}{3} + \frac{1}{11} + \frac{1}{u} \cdots \right) & u > 0 \\ E_5'' - E_0 = D\left(\frac{u}{5} + \frac{1}{33} + \frac{1}{u} \cdots \right) & u < 0. \end{cases}$$

That is: For pronounced tetragonal symmetry it is unimportant that E'_5 and E''_5 are two terms with the same representation of the tetragonal group. One gets two terms which (apart from slight corrections of the 6th order) simply correspond to the two orientations of the electron's angular momentum l = 3 with the components m = 1 and m = 3 in the direction of the tetragonal axis:

$$E_3 - E_0 = \varepsilon_{33} - E_0 = D\left(-\frac{u}{3} + \frac{1}{11} \right) = \frac{e}{3} \cdot \frac{\overline{r^2}}{2!} \left(\frac{\partial^2 V}{\partial z^2} \right)_0 - \frac{e}{11} \frac{\overline{r^4}}{4!} \left(\frac{\partial^4 V}{\partial z^4} \right)_0,$$

$$E_1 - E_0 = \varepsilon_{11} - E_0 = D\left(\frac{u}{5} + \frac{1}{33} \right) = -\frac{e}{5} \cdot \frac{\overline{r^2}}{2!} \left(\frac{\partial^2 V}{\partial z^2} \right)_0 - \frac{e}{33} \frac{\overline{r^4}}{4!} \left(\frac{\partial^4 V}{\partial z^4} \right)_0.$$

But for positive tetragonality E'_5 corresponds to an orientation of the angular momentum with the component m = 1 in the direction of the tetragonal axis, whereas for negative u it corresponds to the orientation with m = 3. Independently of the value of u, E'_5 always remains the higher of the two terms, and one cannot consistently assign to it the same electric quantum number, say m = 3, for all u, because this in one case corresponds to the lower and in the other to the higher term. If, beginning with positive tetragonality, we think of the crystal as being adiabatically deformed and approaching cubic symmetry, then it no longer has any meaning to speak of a component m = 1 in the direction of the tetragonal axis, as soon as this tetragonal axis is no longer sufficiently distinguished from the other two axes. Only when the deformation has proceeded so far beyond the cubic case that the symmetry has again become pronouncedly, but now negatively, tetragonal, can we again define a component of the angular momentum of the electron orbit in the direction of the tetragonal axis, but this is now m = 3. This example can be regarded as an illustration of the general theorem that terms of the same representation never cross over each other. We could, of course, also write the zeroth order eigenfunctions (eq. (7), § 11) in terms of u, and would obtain the same transition from P_3^1 through the eigenfunction for cubic symmetry and on to P_3^3, when we let u go from large positive values through zero to large negative values.

The dependence of the term values, calculated relative to the center of gravity of the terms, on the effective tetragonality u is shown in Figs. 6-8; the term values are plotted in terms of the factor[1] $D = -e\frac{\overline{r^4}}{4!}\frac{\partial^4 V}{\partial z^4}$. For added clarity abscissas are given in terms not only of the effective tetragonality u but also of the ratio of axes c/a, which would correspond to the tetragonality u for a stretched crystal of the NaCl type, if the radius of the electron orbit is taken to be about $^1/_4$ of the lattice spacing.

Fig. 6 shows the splitting pattern of the d term as function of the tetragonality. Since no representation has two terms corresponding to it, every term can have assigned to it an m-value (electric quantum number, component of angular momentum in the direction of the tetragonal axis) and an angular eigenfunction independent of u, and in addition each term value depends linearly on u. For pronounced tetragonal symmetry one gets a wide splitting proportional to $\left(\frac{\text{orbit radius}}{\text{lattice spacing}}\right)^2$ between terms with different values of m, for which the usual interval rule (separation between m and m + 1 proportional to $(m+1)^2 - m^2 = 2m + 1$) holds approximately, and a narrow splitting proportional to $\left(\frac{\text{orbit radius}}{\text{lattice spacing}}\right)^4$ between the two terms with the eigenfunctions

$$\sqrt{2}\,P_2^2 \cos 2\varphi \quad \text{and} \quad \sqrt{2}\,P_2^2 \sin 2\varphi .$$

[1] In reality this factor itself will of course change, though indeed slowly, with the change of the tetragonality, but this change will depend on the special structure of the crystal, while the relative positions and separations of the terms depend only on the value of the tetragonality.

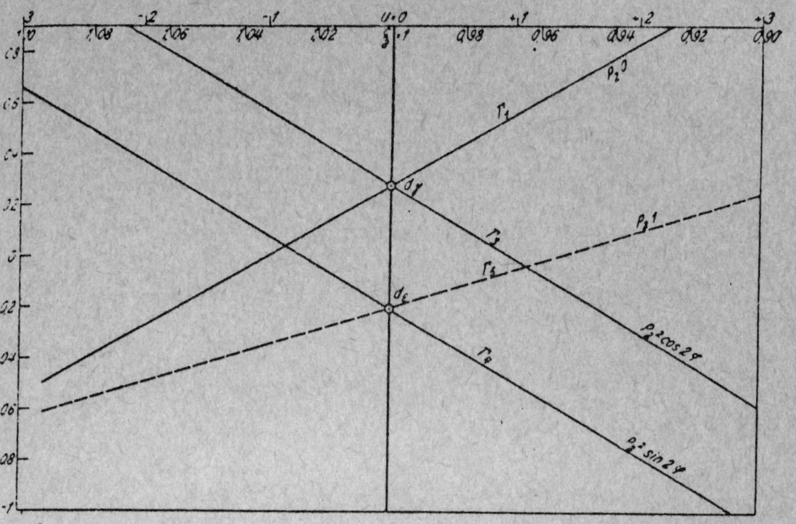

Fig. 6. Splitting pattern of the d term for tetragonal symmetry.

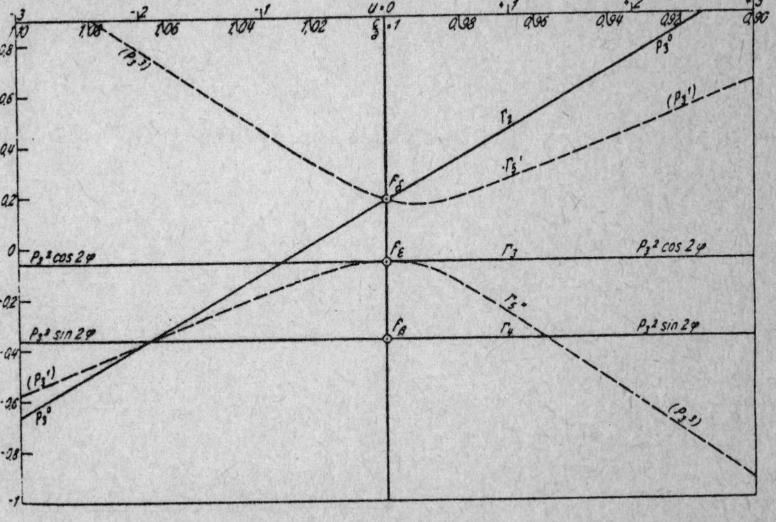

Fig. 7. Splitting pattern of the f term for tetragonal symmetry

– – – – – two-fold degenerate terms
———— simple terms

Fig. 7 shows the splitting pattern of the f term as function of u. The curves for the terms E'_5 and E''_5 form a hyperbola, whose asymptotes are the straight lines $\epsilon_{11} - E_0$ and $\epsilon_{33} - E_0$. These lines would represent the terms belonging to the electric quantum numbers 1 and 3, if these quantum numbers could be defined for all u.

Fig. 8 shows the splitting pattern of the g term, in which four terms are represented as functions of u by hyperbolas. For large tetragonality, one again gets an ordering of the terms according to m; the term m = 2 is split up once again (eigenfunctions $\sqrt{2}\ P_4^2 \cos 2\varphi$ and $\sqrt{2}\ P_4^2 \sin 2\varphi$), but the corresponding splitting

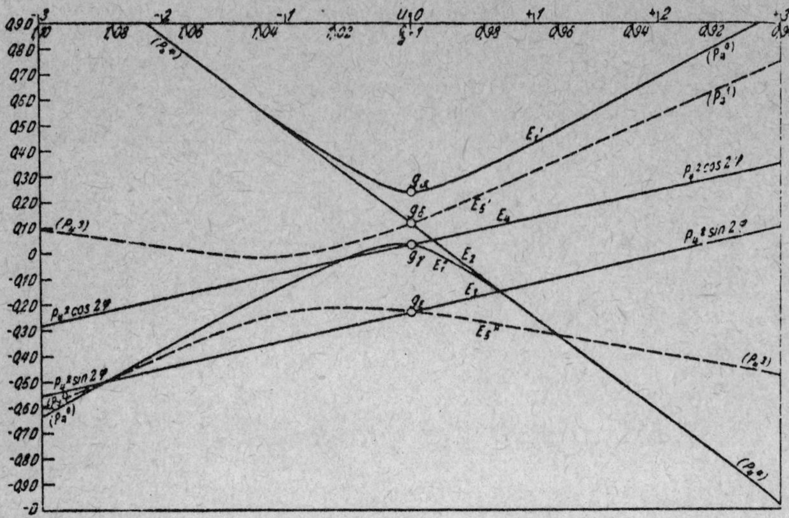

Fig. 8. Splitting pattern of the g term for tetragonal symmetry.

for m = 4 practically vanishes, because it is produced in lowest order by the multipole moment of 8th order.

As regards the absolute magnitude of the splitting, for pronounced tetragonal symmetry, this can reach considerably larger values than for cubic symmetry. If we take, say, a crystal of the shortened NaCl type, then for an axis ratio

$$\frac{c}{a} = \frac{7}{8} \quad \text{one has say} \quad \frac{1}{2!}\left(\frac{\partial^2 V}{\partial z^2}\right)_0 = -1 \cdot \frac{E}{a^3},$$

and thus for an orbit radius of $^1/_5$ the lattice spacing, and for singly charged ions, the term value is:

$$E_\lambda - E_0 \approx + \frac{\overline{r^2}}{a^2} \cdot \frac{e^2}{a} \cdot \alpha_{\lambda\lambda} \approx \frac{1}{25}\,\frac{e^2}{a}\,\alpha_{\lambda\lambda} \approx$$
$$0{,}008 \cdot 2 \cdot \text{Rydberg constant} \cdot \alpha_{\lambda\lambda} = \text{say } 2000\ \text{cm}^{-1}.$$

Thus the splitting can easily exceed 1000 cm^{-1}.

Since the tetragonality controls the relative position of the crystal terms, it determines the most stable arrangement of electrons in the crystal, as has been said before. For example, in the case of strong negative tetragonality, for a d electron of a positive ion the orientation of the electron along the tetragonal axis is the most stable (eigenfunction P_2^0); for less pronounced stretching one finds an energy minimum for an oblique orientation, in which the maximum of the density is located on a cone around the tetragonal axis with aperture angle $\pi/2$ (P_2^1). For cubic symmetry the stablest state has the density maximum along the three-fold axes, and finally for positive tetragonality (shortened NaCl crystal) it goes over onto the face diagonals of the plane perpendicular to the tetragonal axis (eigenfunction $\sqrt{2}\, P_2^2 \sin 2\varphi$). Similar considerations can be carried through by means of Figs. 6-8 for negative ions and for other electronic terms. The relative term separations shown in the diagrams are independent of the special structure of the crystal, as is indeed the whole concept of the tetragonality.

§ 23. Prospects for Applications of the Theory.

This theory gives a survey of the effects of electric fields of definite symmetries on an atom. Accordingly it will be possible to apply it to advantage in the treatment of symmetrical molecules[1]. A direct physical confirmation should be obtainable by analyses of the spectra of crystals; it appears that the salts of the rare earths, which are the only ones giving sharp absorption lines, can serve as verification for the theory[2]. Further more it would be conceivable that when supplemented by a theory of the exchange effects this theory could provide an explanation of polymorphic transitions, but we leave this question to one side for the present. Some importance attaches to the negative result, that for cubic symmetry P terms do not split up (and more generally S terms do not). Finally, it is possible to get a survey of the departures of the atomic symmetry in a crystal from spherical symmetry; a comparison of the value of the term splitting of an electron with quantum numbers (n, l) with its term value in the free atom also provides a quantitative estimate of this departure. For tetravalent carbon (diamond) one should, to be sure, expect no important departure from spherical symmetry in zeroth approximation, since the ground term is a ^4S term, so that only the spin can take different orientations, and not the orbital angular momentum. Presumably in this case, it is the deformation of the L shell that will cause the departure from spherical symmetry that is observed by means of x-rays.

(Received July 20, 1929)

[1]See the remarks in § 12, 13, etc. regarding the relations of the present study to that made by Ehlert for CH_4, loc. cit.

[2]I owe my thanks to Dr. Schutz (Tubingen) for pointing this out.

Note added in proof: The absorption lines in question show a splitting in a magnetic field into only a few (often 2) components, which are often widely separated. If one recalls that the crystalline field produces a considerable splitting-up of the terms, this is easily understandable: The unperturbed crystal terms (without magnetic field) are in general only two-fold degenerate or not degenerate at all, and the magnetic field removes this last remaining directional degeneracy and splits the term into only one or two components. On the other hand the separation of these components is proportional to the magnetic quantum number $\underline{m}$, which can take very sizable values because of the large total angular momenta ($\underline{j}$ up to say 10) of the rare earths. A more thorough analysis of these relations is not possible at present on account of the unperspicuous condition of the experimental information.